MicroRNAs in Development and Cancer

Molecular Medicine and Medicinal Chemistry

Book Series Editors: Professor Colin Fishwick *(School of Chemistry, University of Leeds, UK)*
Dr Paul Ko Ferrigno and Professor Terence Rabbitts FRS, FMedSci *(Leeds Institute of Molecular Medicine, St. James's Hospital, UK)*

Forthcoming:

Merkel Cell Carcinoma: A Multidisciplinary Approach
edited by Vernon K. Sondak, Jane L. Messina, Jonathan S. Zager, and Ronald C. DeConti (H Lee Moffitt Cancer Center & Research Institute, USA)

DNA Deamination and the Immune System: Aid in Health and Disease
edited by Sebastian Fugmann, Marilyn Diaz (National Institute of Health, USA), and Nina Papavasiliou (Rockefeller University, USA)

Fluorine in Pharmaceutical and Medicinal Chemistry: From Biophysical Aspects to Clinical Applications
edited by Véronique Gouverneur (University of Oxford, UK) and Klaus Müller (F Hoffmann-La Roche AG, Switzerland)

Molecular Exploitation of Apoptosis Pathways in Prostate Cancer
by Natasha Kyprianou (University of Kentucky, USA)

Antibody Drug Discovery
edited by Clive R. Wood (Bayer Schering Pharma, Germany)

Volume **1**

**Molecular Medicine and
Medicinal Chemistry**

MicroRNAs in Development and Cancer

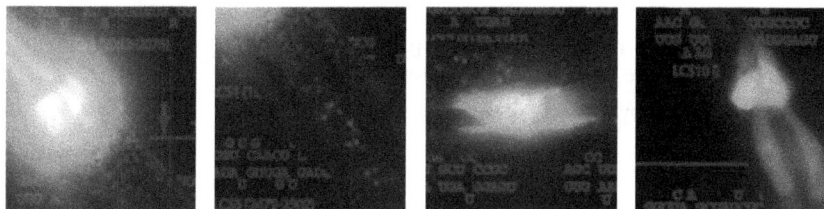

Editor

Frank J Slack

Yale University, USA

Imperial College Press

ICP

Published by

Imperial College Press
57 Shelton Street
Covent Garden
London WC2H 9HE

Distributed by

World Scientific Publishing Co. Pte. Ltd.
5 Toh Tuck Link, Singapore 596224
USA office: 27 Warren Street, Suite 401-402, Hackensack, NJ 07601
UK office: 57 Shelton Street, Covent Garden, London WC2H 9HE

British Library Cataloguing-in-Publication Data
A catalogue record for this book is available from the British Library.

MICRORNAS IN DEVELOPMENT AND CANCER
Molecular Medicine and Medicinal Chemistry — Vol. 1

Copyright © 2011 by Imperial College Press

ISBN-13 978-1-84816-366-9
ISBN-10 1-84816-366-5

Typeset by Stallion Press
Email: enquiries@stallionpress.com

Printed in Singapore.

Contents

Preface

*Frank J. Slack**

.

The human genome encodes hundreds of non-coding RNAs, such as microRNAs (miRNAs). MiRNAs have recently emerged as key regulators of gene expression during development and are frequently mis-expressed in human disease states, in particular cancer. MiRNAs act to promote or repress cell proliferation, migration and apoptosis during development, all of them processes that go awry in cancer. Thus, miRNAs have the ability to behave like oncogenes or tumour suppressors. Their small size and molecular properties make miRNAs amenable as targets and therapeutics in cancer treatment. MiRNAs thus present a paradigm shift in thinking about gene regulation during development and disease, and provide the oncologist with a potentially powerful new battery of agents to diagnose and treat cancer. This book brings together investigators from many distinct fields — namely RNA, bioinformatics, genomics and cancer biology — to facilitate new ideas regarding the means by which non-coding RNAs such as miRNAs impact tumour initiation and progression. The science discussed here has the potential to seed the next generation of anti-cancer diagnostics and therapeutics. While microRNAs have led the charge in the emergence of small RNA molecules functioning during development and

*Department of Molecular, Cellular, and Developmental Biology, Yale University, P.O. Box 208103, 266 Whitney Ave, New Haven, Connecticut, USA. E-mail: frank.slack@yale.edu

disease, other small and large RNAs also contribute, and are also the subject of chapters in this book.

Cancer is a huge worldwide public health issue. The disease has recently become the greatest killer of older Americans, with hundreds of thousands of deaths annually. While oncologists have made progress in the early diagnosis and treatment of cancer, there is still room for improvement in this area. In many cases, novel therapies have emerged from an understanding of the basic biology of the disease. Most studies of the genetic causes of cancer have focused on mutations and alterations in protein-coding genes. However, scientists now appreciate that the genome generates a large diversity of non-coding RNAs, many of which have unknown functions, but which are increasingly implicated in cancer. Accumulating data suggests causal roles for miRNAs in human cancer, including observations of miRNA genes at tumour-associated chromosomal lesions and direct demonstrations that altered expression of miRNAs can cause tumour development. Analysis of miRNA expression signatures has also shown promise in contributing to cancer diagnosis, and miRNAs themselves are potential targets/agents for cancer therapy.

While hundreds of human miRNAs and non-coding RNAs are known, relatively little is understood about their roles and targets, and there is still a limited literature on using these molecules in the clinic. This book will focus on the exciting biology of non-coding RNAs, in particular, miRNAs in controlling developmental and cancer processes like cell proliferation, differentiation, cell cycle, apoptosis and metastasis. This book will highlight the best current research into the roles that miRNAs play in these fundamental processes and will provide the basic understanding that is driving the invention of powerful clinical tools.

Given the sheer number of miRNA genes, approaching 1,000 in humans, and our lack of understanding of all their roles, and given the need for a basic understanding of cancer mechanisms, it is hoped that this book will provide an important synthesis of the growing appreciation for miRNAs in development and cancer. It will highlight emerging roles of miRNAs in tumourigenesis and place these in the context of the normal roles for miRNAs in development and homeostasis. The book will involve chapters from the pre-eminent miRNA and non-coding RNA specialists, with discussions ranging from genomics of natural miRNA genes, to normal roles for miRNAs and the consequences of their loss, to uses of miRNAs in the diagnosis and therapy of cancer. Non-coding RNAs, like miRNAs, have generated much excitement not only because they are intricately involved in cancer, but because they provide druggable targets as well as small molecule therapeutics that could see widespread use in treating cancer. This book will also focus on the latest efforts to harness the power of these small RNAs as agents in the fight against cancer. Discussions will range from the use of miRNAs as diagnostic and prognostic markers in cancer, to single nucleotide polymorphisms (SNPs) providing risk factors for cancer, to the delivery and effectiveness of small RNA therapeutics in clinically relevant settings. This book will advance the field and accelerate the benchside to bedside use of this technology.

The confluence of the superb group of basic and translational scientists assembled in this book is highly likely to stimulate exciting discussion and drive future discoveries in this important area.

1

Introduction to Gene Regulation by Small RNAs

*Allison L. Abbott**

Gene regulation by small RNAs is now recognised to be critical in animal development and physiology. Human disease conditions including cancer and cardiovascular disease are often associated with the misregulation of microRNAs. MicroRNAs (miRNAs) are ~22 nucleotide non-coding RNAs that are generated by the activity of the RNase III enzyme, Dicer. The biologically relevant target messenger RNAs (mRNAs) for the majority of microRNAs remain unknown. Target mRNAs are controlled by microRNAs at the step of translation initiation or elongation. Piwi-interacting RNAs (piRNAs) are ~24–30 nucleotide non-coding RNAs that are generated by a Dicer-independent mechanism and function in the germline to prevent the spreading of mobile genetic elements.

1.1 Introduction

Once considered the sole province of proteins, control of gene expression by a small RNA was first described in 1993 when Victor Ambros and

*Department of Biological Sciences, Marquette University, P.O. Box 1881, Milwaukee, Wisconsin, USA. E-mail: allison.abbott@marquette.edu.

1

co-workers identified the ~22 nucleotide small RNA encoded by the *lin-4* gene in *C. elegans*.[1] This early research defined many of the key elements of post-transcriptional regulation of gene expression by microRNAs. From genetic and molecular evidence, it was clear that *lin-4* represses *lin-14* activity. Therefore, when *lin-4* was found to have antisense complementarity to sequence motifs in the *lin-14* 3′ untranslated region (UTR),[2] the model of a miRNA repressing target gene activity through interaction with the 3′ UTR of a target mRNA was established. However, because no homologues of *lin-4* were identified in higher organisms, it was unclear how much of this regulatory mechanism was conserved. The *lin-4* small RNA remained an oddity until 2000, when the second gene encoding a small regulatory RNA, *let-7*, was identified, first in worms[3] and subsequently in a wide array of species including flies, fish, mice and humans.[4] There now exist several classes of endogenous small RNA regulators of gene expression, including miRNAs as well as piwi-associated small RNAs (piRNAs). Thus, *lin-4* is now viewed as an 'emissary' for a newly-appreciated class of regulatory molecules[5] rather than the 'deviant' it may have once been considered.

1.2 Gene Regulation by MicroRNAs

A miRNA is defined as a small, single-stranded RNA, 19–24 nucleotides in length that is generated following the cleavage of a longer precursor transcript by the RNase III-type enzyme Dicer. Loci encoding miRNAs are found in the genome in independent transcription units, in clusters of multiple miRNAs, or within introns of annotated genes.[6] Experimentally, miRNAs have been identified by sequencing complementary DNA (cDNA) libraries derived from size-selected RNA samples or by computational predictions. These candidate miRNAs are then validated by using detection methods such as northern blotting, RT-PCR, or microarray analysis.[7] These methods have led to the identification of thousands of miRNAs in animals, plants and viruses. Release 15 of the miRBase database lists 14,197 entries for miRNAs. The identification of miRNAs has thus far vastly outpaced miRNA functional analysis.

These miRNAs are generated in a cell following multiple processing steps in both the nucleus and the cytoplasm to generate the ~22 nt mature miRNA.[8] Unlike other small RNAs such as transfer RNAs or U6 small nuclear RNA, most miRNAs are transcribed by RNA Polymerase II (Pol II) as long RNA precursors called primary miRNAs (pri-miRNA). Like other Pol II transcripts, pri-miRNAs are capped and polyadenylated.[9,10] Pri-miRNAs undergo co-transcriptional processing mediated by the Microprocessor complex, which includes the RNase III enzyme Drosha and DGCR8/Pasha, in order to generate a ~70 nucleotide stem-loop precursor miRNA (pre-miRNA).[11,12] Those miRNAs that reside in the introns of annotated genes can be processed either by the Drosha complex before the splicing reaction or by the pre-mRNA splicing machinery to generate the pre-miRNA.[13–15] This ~70 nt stem-loop pre-miRNA is exported from the nucleus into the cytoplasm where it is processed by another RNase III enzyme, Dicer.[16–20] The mature, single-stranded miRNA functions in a protein complex called a miRISC (RNA-induced silencing complex), to direct the repression of target mRNAs.

1.2.1 *Mechanisms for microRNA-mediated regulation of gene expression*

The function of a miRNA contained in a miRISC is to identify target mRNAs through sequence complementarity in order to block protein production. An Argonaute protein is a core constituent of the miRISC and is necessary for miRNA activity. Flies that lack activity of the Argonaute-encoding gene, *dAGO1*, die during embryogenesis or early larval development and display defects in miRNA maturation or stability.[21,22] Similarly, worms that lack activity of two Argonaute-encoding genes, *alg-1* and *alg-2*, have reduced miRNA levels and die early in development.[17] Although worms have at least 26 genes encoding Argonaute proteins, *alg-1* and *alg-2* are the only ones that have been found to function in the miRNA pathway.[17,23] Argonaute proteins contain a PAZ domain that can recognise miRNAs by their characteristic Dicer-generated 3′ overhangs as well as a PIWI domain that is important for target repression by miRNAs.[24,25] Nearly all miRNAs in animals bind to their mRNA targets with imperfect complementarity. This imperfect complementarity in binding results in bulges present in the miRNA–mRNA duplex; this precludes the specific Argonaute-mediated cleavage known as Slicer activity and instead leads to target mRNA repression of translation. In contrast, in the RNA interference (RNAi) pathway, Slicer activity cleaves mRNA targeted by a perfectly matched small interfering RNA (siRNA).

While it is clear that nearly all miRNA targets are not specifically cleaved by Slicer activity, the exact mechanism for post-transcriptional repression is not yet fully understood and remains controversial.[26] It may be that multiple mechanisms for miRNA-mediated repression exist and the fates of individual target mRNAs will vary. Early studies demonstrated that the *lin-4* miRNA directed the dramatic reduction of LIN-14 and LIN-28 protein levels while effecting little change on their mRNA levels. Target mRNAs co-fractionate with polyribosomes along with miRNAs.[27,28] These polysomes are sensitive to puromycin indicating that they are in the process of active elongation.[29–31] This result suggested that miRNA-mediated repression can occur after the initiation of translation. Inhibition of translation can occur post-initiation due to a block in translation elongation or due to degradation of the nascent polypeptide chain.[30,31]

In addition to blocking translation at a step post-initiation, miRNAs can direct the degradation of the target mRNA, due to decapping or deadenylation. For example, miR-430 is expressed in zebrafish during early embryogenesis and mediates the deadenylation and loss of maternal mRNA targets at the maternal-to-zygotic transition.[32] Deadenylation and degradation of target mRNAs is also seen in flies and in mammalian cells.[33,34] One model is that miRNAs function to direct target mRNAs towards processing bodies, or P-bodies, that contain enzymes involved in RNA decay for storage or for degradation.[35,36]

However, more recent studies suggest that miRNA-mediated repression of target mRNAs can also occur at the step of translation initiation. First, in cultured mammalian cells, a target mRNA reporter shifts towards the lighter fractions of a sucrose gradient when repressed by a miRNA, suggesting a block of translation initiation.[37] In contrast, as discussed above, *lin-14* and *lin-4* are found in the heavier, polyribosome-containing fractions, suggesting a block after initiation. Secondly, reporter mRNAs under the regulation of viral internal ribosome entry sites (IRES) do not appear to be repressed by miRNAs, suggesting that recognition of the 5′ cap of a target mRNA may be required for miRNA-mediated repression.[37,38] Thirdly, human Ago2 is able to associate with the m^{7}G structure and this association is necessary for translational repression.[39] Lastly, in order to determine the mechanism for repression, proteins that function in the miRISC have been identified. Factors of the core translational initiation complex such as polyadenylate-binding proteins, an eIF4G orthologue and a component of the 40S ribosome have been demonstrated to associate with the miRISC factors, AIN-1 and AIN-2.[40] Notably, proteins in the decapping complex were not identified.[40] Thus, one model for miRISC activity is that proteins of the miRISC bind to the cap of target mRNAs and block access of key initiation factors to the target mRNA.

These models are not mutually exclusive, as there is evidence to support multiple mechanisms for miRNA-mediated repression of target mRNAs. One possibility is that miRNAs may effect these different sets of changes on their targets dependent upon other regulatory proteins associated with a miRISC or other modifications to the miRISC or to the target mRNA. In support of this, it was observed that the promoter region of a target could govern the type of miRNA-mediated regulation. Those mRNAs under the regulation of an SV40 promoter are repressed at the initiation step whereas those mRNAs under the regulation of a TK promoter are repressed after the initiation step, as determined by sucrose gradient analysis.[41] Factors that associate with the target mRNAs in order to determine the mechanism by which a target is repressed have not been identified.

1.2.2 *Identification of microRNA-regulated target mRNAs*

In animals, miRNAs bind to their target mRNAs with imperfect sequence complementarity. As stated above, functionally, this imperfect binding precludes Slicer-mediated mRNA cleavage of target mRNAs. Computationally, this imperfect binding precludes direct identification of target mRNAs by simple analysis of the nucleotide sequence of 3′ UTRs. There have been various strategies employed to identify candidate target mRNAs for individual miRNAs in many species. One strategy is to identify 3′ UTRs that have binding sites with high complementarity to the 7–8 nucleotide 'seed' sequence at the 5′ end of miRNAs and that are conserved across closely related species. Such approaches have identified thousands of possible miRNA target mRNAs in humans, flies, worms and other animals.[42–46] In addition, surrounding sequence motifs around binding sites,[47] compensatory binding of the 3′ end of the miRNA,[42] combinatorial control of a mRNA by multiple miRNAs[44] and secondary structure of the 3′ UTRs of targets[48,49] have been incorporated into target prediction algorithms and may improve the accuracy of predictions.

However, while these predictions are highly useful tools to begin to understand the possible functions of specific miRNAs, it is expected that these lists contain many false positives and may fail to identify biologically relevant targets that do not fit the current models of miRNA–mRNA binding. In addition to computational approaches, biochemical approaches have been taken to identify biologically relevant miRNA targets.[40,50–53] Combining these bioinformatic and biochemical data sets[54] is expected to greatly strengthen the ability to accurately predict target mRNAs for individual miRNAs.

The biological significance of miRNA-mediated gene regulation has been characterised by analysis of mRNA and protein levels following changes in miRNA activity. Transfection of a single miRNA into cultured cells results in modest changes in mRNAs levels, many of which are predicted to be direct targets of the transfected miRNA.[55] In addition to analysis of changes at the mRNA level, analysis of changes at the protein level can also be used to determine the function of specific miRNAs. Recent advances in proteomics approaches were used to determine the effect of introducing a single miRNA to cultured cells or inhibiting the activity of an endogenous miRNA.[56,57] Two important themes emerge from these studies: first, the impact of miRNA regulation on an individual gene's activity is modest, with most changes at the two-fold level, and second, hundreds of genes displayed change in protein expression. Therefore, as these cases of mis-expressed miRNA activity suggest, some miRNAs function to modulate the overall protein production in a cell rather than strongly repress a specific subset of targets.

An important function for some miRNAs is to fine-tune the activity of a suite of target mRNAs, acting as 'micromanagers' of gene expression.[58] The ability of a single miRNA to bind a target mRNA with only seven nucleotides of complementarity can allow for the repression of a large number of genes, which may globally influence the protein production of a particular cell. For example, introduction of the brain-specific miRNA, miR-124, into HeLa cells results in changes in mRNA levels of 174 genes marking a global shift in the expression profile of HeLa cells to more closely resemble the profile of brain tissue.[55] Thus, miR-124 may act to promote or stabilise tissue-specific identity in the brain through modest repression of a suite of targets. In addition, miR-430 and miR-309-cluster miRNAs in fish and flies, respectively, regulate a suite of target mRNAs at the maternal-zygotic transition in order to remove hundreds of maternal transcripts from the early embryo.[32,59]

In contrast to examples of individual miRNAs that regulate a large set of targets, other miRNAs appear to fine-tune, or modulate, the expression of only a few key mRNA targets. These miRNAs can function to subtly repress targets in order to keep target protein levels within an optimal range. In these cases, it is expected that miRNA and target are co-expressed and also that both too little and too much target protein would cause abnormalities. One example of this type of miRNA-target relationship is miR-9a regulation of *senseless* activity during sensory bristle development in flies.[60] In the absence of miR-9a, there is a small and variable increase in the number of sensory bristles. Mir-9a regulation may function to keep *senseless* activity below a threshold in those cells not fated to become sensory bristles. While transcriptional control can be sufficient, the additional post-transcriptional control provided by miR-9a confers robustness on this developmental pathway and ensures that no ectopic bristles are specified.

While some miRNAs that are co-expressed with target mRNAs may act to modulate protein levels or act in conjunction with transcriptional control of targets, other miRNAs function to fully repress the activity of their targets. Only a few of these so-called 'switch' targets have been identified to date. These would be the most readily identified by genetic screens. Indeed, the first miRNAs identified, *lin-4* and *let-7*, were found using forward genetics and act to nearly completely repress the target mRNAs, *lin-14* and *lin-41*, respectively.[61,62] Mutations in these genes can suppress the strong, penetrant phenotypes associated with loss of *lin-4* or *let-7*. In flies, the *let-7* miRNA regulation of neuromuscular junction development is possibly mediated through the control of the *abrupt* gene.[63,64] Thus, in several cases the functions of miRNAs are mediated primarily through strong repression of single targets or a small number of targets. However, although there are a large number of miRNA-encoding genes that have been identified in flies, fish and worms, only a tiny fraction of these have been found using forward genetic screens. Although this could partly be due to the relatively small target size of miRNA-encoding loci, it suggests that 'switch' targets that are essential for developmental pathways are rare. Whether they are fine-tuning or switch targets, the identification of the biologically relevant targets controlled by specific miRNAs to regulate animal development or physiology is a crucial challenge in the field.

1.3 Gene Regulation by piRNAs

In 2006, a new class of RNA regulatory molecules of about ~26–32 nucleotides in length, the Piwi-interacting RNAs (piRNAs), was identified.[65–69] There are many differences between piRNAs and miRNAs. Firstly, the size of piRNAs is greater than the 21–23 nucleotide miRNAs, indicating a different biogenesis pathway. Indeed, production of piRNAs does not require Dicer activity.[70,71] Secondly, piRNAs are most robustly expressed in the germline, whereas miRNAs are more broadly expressed. Thirdly, most piRNAs are generated from a small number of distinct loci termed piRNA clusters that are largely composed of transposon and repetitive sequences;[72] in contrast, genes encoding miRNAs are found dispersed in animal genomes.

1.3.1 Mechanism for piRNA-mediated regulation of gene expression

Evidence supports a model in which piRNAs are processed from a single-stranded RNA precursor transcript. Because the transposon sequences in piRNA clusters are incomplete or damaged, they are not likely to be able to be expressed or mobilised. Like miRNAs, piRNAs interact and function with a protein in the Argonaute/Piwi (P-element-induced wimpy testis) family. Whereas Argonaute proteins in the miRISC are not catalytically active, all three Piwi proteins in flies (Piwi, Aubergine and Ago3) have Slicer activity and direct the cleavage of target RNAs between nucleotides 10 and 11.[73,74] Ago3-bound piRNAs are derived from the sense strand of transposons, whereas Piwi and Aubergine-bound piRNAs are derived

from the antisense strand.[72,73] After the processing of the primary transcript that is derived from a piRNA cluster, there is an amplification loop.
Active transposon transcripts are recognised by antisense strand piRNA-
containing Piwi/Aubergine complexes and cleaved, which after a second 3′
cleavage event generates additional sense strand piRNAs that can then
associate with Ago3. This self-reinforcing amplification loop has been
termed the 'ping-pong model'.[72,73]

Slicer activity by Piwi proteins generates the 5′ end of piRNAs.
Additional proteins are necessary for cleavage as well as for methylation at
the 3′ end of mature piRNAs. In flies, nucleases encoded by *zucchini* and
squash genes appear to function in piRNA biogenesis as mutations in these
genes result in reduced piRNA production and silencing of retrotransposons.[75] The RNA methyltransferase in flies, DmHen1/Pimet, functions
to methylate the 3′ end of piRNAs.[76,77] However, unlike Piwi mutants,
DmHen1/Pimet mutants are viable, indicating that this methylation event
is not essential for piRNA activity.[77]

The silencing of transposons is essential in the germline where the
Piwi subfamily is primarily expressed in flies and mice,[78] although it is
likely that additional functions exist for piRNAs in the soma.[79–81] The Piwi
protein was first described in 1997 in studies of germline development in
flies, in which Piwi is necessary to maintain germline stem cells.[82] Defects
in the piRNA pathway results in overexpression of retrotransposons and
DNA damage.[78] In mice, the three Piwi orthologues, MIWI, MILI and
MIWI2, are all expressed in the testes and are required for spermatogenesis.[78]
Mutations in any of the Piwi orthologues results in DNA damage and the
induction of the apoptosis pathway in germline cells.[83–85] DNA damage is
likely due to a failure to silence transposons,[86] suggesting a conserved
function of Piwi proteins and piRNAs in blocking DNA damage due to
retrotransposon activity.

1.4 Summary and Perspectives

Even with the astonishing progress made in the last decade of research, the
landscape of small RNA regulation of gene expression is still expanding.
The biological activity of the majority of miRNAs and their critical target
mRNAs are largely unknown. Chapters in this volume provide detailed

analysis of the current state of our knowledge of the functions of miRNAs and piRNAs during animal development and in the initiation or progression of human cancers.

Acknowledgments

A.L. Abbott is supported by NIH/NIGMS grant GM-084451.

References

1. Lee R.C., Feinbaum R.L. and Ambros V. (1993). The *C. elegans* heterochronic gene *lin-4* encodes small RNAs with antisense complementarity to *lin-14*. *Cell* **75**, 843–854.
2. Wightman B., Ha I. and Ruvkun G. (1993). Posttranscriptional regulation of the heterochronic gene *lin-14* by *lin-4* mediates temporal pattern formation in *C. elegans. Cell* **75**, 855–862.
3. Reinhart B.J., Slack F.J., Basson M. *et al.* (2000). The 21-nucleotide *let-7* RNA regulates developmental timing in *Caenorhabditis elegans. Nature* **403**, 901–906.
4. Pasquinelli A.E., Reinhart B.J., Slack F. *et al.* (2000). Conservation of the sequence and temporal expression of *let-7* heterochronic regulatory RNA. *Nature* **408**, 86–89.
5. Wickens M. and Takayama K. (1994). Deviants — Or emissaries? *Nature* **367**, 17–18.
6. Bartel D.P. (2004). MicroRNAs: Genomics, biogenesis, mechanism, and function. *Cell* **116**, 281–297.
7. Ambros V., Bartel B., Bartel D.P. *et al.* (2003). A uniform system for microRNA annotation. *RNA* **9**, 277–279.
8. Kim V.N. (2005). MicroRNA biogenesis: Coordinated cropping and dicing. *Nat Rev Mol Cell Biol* **6**, 376–385.
9. Cai X., Hagedorn C.H. and Cullen B.R. (2004). Human microRNAs are processed from capped, polyadenylated transcripts that can also function as mRNAs. *RNA* **10**, 1957–1966.
10. Lee Y., Kim M., Han J. *et al.* (2004). MicroRNA genes are transcribed by RNA polymerase II. *EMBO J* **23**, 4051–4060.
11. Lee Y., Ahn C., Han J. *et al.* (2003). The nuclear RNase III Drosha initiates microRNA processing. *Nature* **425**, 415–419.

12. Morlando M., Ballarino M., Gromak N. *et al.* (2008). Primary microRNA transcripts are processed co-transcriptionally. *Nat Struct Mol Biol* **15**, 902–909.

13. Kim Y. and Kim V.N. (2007). Processing of intronic microRNAs. *EMBO J* **26**, 775–783.

14. Okamura K., Hagen J.W., Duan H. *et al.* (2007). The mirtron pathway generates microRNA-class regulatory RNAs in *Drosophila. Cell* **130**, 89–100.

15. Ruby J.G., Jan C.H. and Bartel D.P. (2007). Intronic microRNA precursors that bypass Drosha processing. *Nature* **448**, 83–86.

16. Bernstein E., Caudy A.A., Hammond S.M. *et al.* (2001). Role for a bidentate ribonuclease in the initiation step of RNA interference. *Nature* **409**, 363–366.

17. Grishok A., Pasquinelli A.E., Conte D. *et al.* (2001). Genes and mechanisms related to RNA interference regulate expression of the small temporal RNAs that control *C. elegans* developmental timing. *Cell* **106**, 23–34.

18. Hutvágner G., McLachlan J., Pasquinelli A.E. *et al.* (2001). A cellular function for the RNA-interference enzyme Dicer in the maturation of the *let-7* small temporal RNA. *Science* **293**, 834–838.

19. Ketting R.F., Fischer S.E., Bernstein E. *et al.* (2001). Dicer functions in RNA interference and in synthesis of small RNA involved in developmental timing in *C. elegans. Genes Dev* **15**, 2654–2659.

20. Knight S.W. and Bass B.L. (2001). A role for the RNase III enzyme DCR-1 in RNA interference and germ line development in *Caenorhabditis elegans. Science* **293**, 2269–2271.

21. Kataoka Y., Takeichi M. and Uemura T. (2001). Developmental roles and molecular characterization of a *Drosophila* homologue of *Arabidopsis Argonaute1*, the founder of a novel gene superfamily. *Genes Cells* **6**, 313–325.

22. Okamura K. (2004). Distinct roles for Argonaute proteins in small RNA-directed RNA cleavage pathways. *Genes Dev* **18**, 1655–1666.

23. Yigit E., Batista P.J., Bei Y. *et al.* (2006). Analysis of the *C. elegans* Argonaute family reveals that distinct Argonautes act sequentially during RNAi. *Cell* **127**, 747–757.

24. Hutvagner G. and Simard M.J. (2008). Argonaute proteins: Key players in RNA silencing. *Nat Rev Mol Cell Biol* **9**, 22–32.

25. Meister G. and Tuschl T. (2004). Mechanisms of gene silencing by double-stranded RNA. *Nature* **431**, 343–349.

26. Liu J. (2008). Control of protein synthesis and mRNA degradation by microRNAs. *Curr Opin Cell Biol* **20**, 214–221.

27. Olsen P.H. and Ambros V. (1999). The *lin-4* regulatory RNA controls developmental timing in *Caenorhabditis elegans* by blocking LIN-14 protein synthesis after the initiation of translation. *Dev Biol* **216**, 671–680.

28. Seggerson K., Tang L. and Moss E.G. (2002). Two genetic circuits repress the *Caenorhabditis elegans* heterochronic gene *lin-28* after translation initiation. *Dev Biol* **243**, 215–225.

29. Maroney P.A., Yu Y., Fisher J. *et al.* (2006). Evidence that microRNAs are associated with translating messenger RNAs in human cells. *Nat Struct Mol Biol* **13**, 1102–1107.

30. Nottrott S., Simard M.J. and Richter J.D. (2006). Human let-7a miRNA blocks protein production on actively translating polyribosomes. *Nat Struct Mol Biol* **13**, 1108–1114.

31. Petersen C.P., Bordeleau M.E., Pelletier J. *et al.* (2006). Short RNAs repress translation after initiation in mammalian cells. *Mol Cell* **21**, 533–542.

32. Giraldez A.J., Mishima Y., Rihel J. *et al.* (2006). Zebrafish MiR-430 promotes deadenylation and clearance of maternal mRNAs. *Science* **312**, 75–79.

33. Behm-Ansmant I., Rehwinkel J., Doerks T. *et al.* (2006). mRNA degradation by miRNAs and GW182 requires both CCR4:NOT deadenylase and DCP1:DCP2 decapping complexes. *Genes Dev* **20**, 1885–1898.

34. Wu L., Fan J. and Belasco J.G. (2006). MicroRNAs direct rapid deadenylation of mRNA. *Proc Natl Acad Sci USA* **103**, 4034–4039.

35. Chan S.P. and Slack F.J. (2006). MicroRNA-mediated silencing inside P-bodies. *RNA Biol* **3**, 97–100.

36. Jakymiw A., Pauley K.M., Li S. *et al.* (2007). The role of GW/P-bodies in RNA processing and silencing. *J Cell Sci* **120**, 1317–1323.

37. Pillai R.S., Bhattacharyya S.N., Artus C.G. *et al.* (2005). Inhibition of translational initiation by *Let-7* MicroRNA in human cells. *Science* **309**, 1573–1576.

38. Humphreys D.T., Westman B.J., Martin D.I. *et al.* (2005). MicroRNAs control translation initiation by inhibiting eukaryotic initiation factor 4E/cap and poly(A) tail function. *Proc Natl Acad Sci USA* **102**, 16961–16966.

39. Kiriakidou M., Tan G.S., Lamprinaki S. *et al.* (2007). A mRNA m7G cap binding-like motif within human Ago2 represses translation. *Cell* **129**, 1141–1151.

40. Zhang L., Ding L., Cheung T.H. *et al.* (2007). Systematic identification of *C. elegans* miRISC proteins, miRNAs, and mRNA targets by their interactions with GW182 proteins AIN-1 and AIN-2. *Mol Cell* **28**, 598–613.

41. Kong Y.W., Cannell I.G., de Moor C.H. *et al.* (2008). The mechanism of micro-RNA-mediated translation repression is determined by the promoter of the target gene. *Proc Natl Acad Sci USA* **105**, 8866–8871.

42. Brennecke J., Stark A., Russell R.B. *et al.* (2005). Principles of microRNA-target recognition. *PLoS Biol* 3, 404–418.

43. Farh K.K., Grimson A., Jan C. *et al.* (2005). The widespread impact of mammalian microRNAs on mRNA repression and evolution. *Science* 310, 1817–1821.

44. Krek A., Grün D., Poy M.N. *et al.* (2005). Combinatorial microRNA target predictions. *Nat Genet* 37, 495–500.

45. Lewis B.P., Burge C.B. and Bartel D.P. (2005). Conserved seed pairing, often flanked by adenosines, indicates that thousands of human genes are microRNA targets. *Cell* 120, 15–20.

46. Lewis B.P., Shih I.H., Jones-Rhoades M.W. *et al.* (2003). Prediction of mammalian microRNA targets. *Cell* 115, 787–798.

47. Grimson A., Farh K.K., Johnston W.K. *et al.* (2007). MicroRNA targeting specificity in mammals: Determinants beyond seed pairing. *Mol Cell* 27, 91–105.

48. Kertesz M., Iovino N., Unnerstall U. *et al.* (2007). The role of site accessibility in microRNA target recognition. *Nat Genet* 39, 1278–1284.

49. Long D., Lee R., Williams P. *et al.* (2007). Potent effect of target structure on microRNA function. *Nat Struct Mol Biol* 14, 287–294.

50. Beitzinger M., Peters L., Zhu J.Y. *et al.* (2007). Identification of human microRNA targets from isolated Argonaute protein complexes. *RNA Biol* 4, 76–84.

51. Easow G., Teleman A.A. and Cohen S.M. (2007). Isolation of microRNA targets by miRNP immunopurification. *RNA* 13, 1198–1204.

52. Hendrickson D.G., Hogan D.J., Herschlag D. *et al.* (2008). Systematic identification of mRNAs recruited to Argonaute 2 by specific microRNAs and corresponding changes in transcript abundance. *PLoS ONE* 3, e2126.

53. Karginov F.V., Conaco C., Xuan Z. *et al.* (2007). A biochemical approach to identifying microRNA targets. *Proc Natl Acad Sci USA* 104, 19291–19296.

54. Hammell M., Long D., Zhang L. *et al.* (2008). mirWIP: MicroRNA target prediction based on microRNA-containing ribonucleoprotein-enriched transcripts. *Nat Methods* 5, 813–819.

55. Lim L.P., Lau N.C., Garrett-Engele P. *et al.* (2005). Microarray analysis shows that some microRNAs downregulate large numbers of target mRNAs. *Nature* 433, 769–773.

56. Baek D., Villen J., Shin C. *et al.* (2008). The impact of microRNAs on protein output. *Nature* 455, 64–71.

57. Selbach M., Schwanhäusser B., Thierfelder N. *et al.* (2008). Widespread changes in protein synthesis induced by microRNAs. *Nature* 455, 58–63.

58. Bartel D.P. and Chen C.Z. (2004). Micromanagers of gene expression: The potentially widespread influence of metazoan microRNAs. *Nat Rev Genet* **5**, 396–400.

59. Bushati N., Stark A., Brennecke J. *et al.* (2008). Temporal reciprocity of miRNAs and their targets during the maternal-to-zygotic transition in Drosophila. *Curr Biol* **18**, 501–506.

60. Li Y., Wang F., Lee J.A. *et al.* (2006). MicroRNA-9a ensures the precise specification of sensory organ precursors in Drosophila. *Genes Dev* **20**, 2793–2805.

61. Ambros V. (1989). A hierarchy of regulatory genes controls a larva-to-adult developmental switch in *C. elegans. Cell* **57**, 49–57.

62. Slack F.J., Basson M., Liu Z. *et al.* (2000). The *lin-41* RBCC gene acts in the *C. elegans* heterochronic pathway between the *let-7* regulatory RNA and the LIN-29 transcription factor. *Mol Cell* **5**, 659–669.

63. Caygill E.E. and Johnston L.A. (2008). Temporal regulation of metamorphic processes in Drosophila by the *let-7* and miR-125 heterochronic microRNAs. *Curr Biol* **18**, 943–950.

64. Sokol N.S., Xu P., Jan Y.N. *et al.* (2008). Drosophila *let-7* microRNA is required for remodeling of the neuromusculature during metamorphosis. *Genes Dev* **22**, 1591–1596.

65. Aravin A., Gaidatzis D., Pfeffer S. *et al.* (2006). A novel class of small RNAs bind to MILI protein in mouse testes. *Nature* **442**, 203–207.

66. Girard A., Sachidanandam R., Hannon G.J. *et al.* (2006). A germline-specific class of small RNAs binds mammalian Piwi proteins. *Nature* **442**, 199–202.

67. Grivna S.T., Pyhtila B. and Lin H. (2006). MIWI associates with translational machinery and PIWI-interacting RNAs (piRNAs) in regulating spermatogenesis. *Proc Natl Acad Sci USA* **103**, 13415–13420.

68. Lau N.C., Seto A.G., Kim J. *et al.* (2006). Characterization of the piRNA complex from rat testes. *Science* **313**, 363–367.

69. Watanabe T., Takeda A., Tsukiyama T. *et al.* (2006). Identification and characterization of two novel classes of small RNAs in the mouse germline: Retrotransposon-derived siRNAs in oocytes and germline small RNAs in testes. *Genes Dev* **20**, 1732–1743.

70. Houwing S., Kamminga L.M., Berezikov E. *et al.* (2007). A role for Piwi and piRNAs in germ cell maintenance and transposon silencing in zebrafish. *Cell* **129**, 69–82.

71. Vagin V.V., Sigova A., Li C. *et al.* (2006). A distinct small RNA pathway silences selfish genetic elements in the germline. *Science* **313**, 320–324.

72. Brennecke J., Aravin A.A., Stark A. *et al.* (2007). Discrete small RNA-generating loci as master regulators of transposon activity in Drosophila. *Cell* **128**, 1089–1103.

73. Gunawardane L.S., Saito K., Nishida K.M. *et al.* (2007). A slicer-mediated mechanism for repeat-associated siRNA 5′ end formation in *Drosophila*. *Science* **315**, 1587–1590.

74. Saito K., Nishida K.M., Mori T. *et al.* (2006). Specific association of Piwi with rasiRNAs derived from retrotransposon and heterochromatic regions in the *Drosophila* genome. *Genes Dev* **20**, 2214–2222.

75. Pane A., Wehr K. and Schüpbach T. (2007). Zucchini and squash encode two putative nucleases required for rasiRNA production in the Drosophila germline. *Dev Cell* **12**, 851–862.

76. Horwich M.D., Li C., Matranga C. *et al.* (2007). The Drosophila RNA methyltransferase, DmHen1, modifies germline piRNAs and single-stranded siRNAs in RISC. *Curr Biol* **17**, 1265–1272.

77. Saito K., Sakaguchi Y., Suzuki T. *et al.* (2007). Pimet, the Drosophila homolog of HEN1, mediates 2′-O-methylation of Piwi- interacting RNAs at their 3′ ends. *Genes Dev* **21**, 1603–1608.

78. Klattenhoff C. and Theurkauf W. (2008). Biogenesis and germline functions of piRNAs. *Development* **135**, 3–9.

79. Grimaud C., Bantignies F., Pal-Bhadra M. *et al.* (2006). RNAi components are required for nuclear clustering of Polycomb group response elements. *Cell* **124**, 957–971.

80. Pal-Bhadra M., Bhadra U. and Birchler J.A. (2002). RNAi related mechanisms affect both transcriptional and posttranscriptional transgene silencing in Drosophila. *Mol Cell* **9**, 315–327.

81. Pal-Bhadra M., Leibovitch B.A., Gandhi S.G. *et al.* (2004). Heterochromatic silencing and HP1 localization in Drosophila are dependent on the RNAi machinery. *Science* **303**, 669–672.

82. Cox D.N., Chao A. and Lin H. (2000). Piwi encodes a nucleoplasmic factor whose activity modulates the number and division rate of germline stem cells. *Development* **127**, 503–514.

83. Carmell M.A., Girard A., van de Kant H.J. *et al.* (2007). MIWI2 is essential for spermatogenesis and repression of transposons in the mouse male germline. *Dev Cell* **12**, 503–514.

84. Deng W. and Lin H. (2002). Miwi, a murine homolog of piwi, encodes a cytoplasmic protein essential for spermatogenesis. *Dev Cell* **2**, 819–830.

85. Kuramochi-Miyagawa S., Kimura T., Ijiri T.W. *et al.* (2004). Mili, a mammalian member of piwi family gene, is essential for spermatogenesis. *Development* **131**, 839–849.

86. Aravin A.A., Sachidanandam R., Girard A. *et al.* (2007). Developmentally regulated piRNA clusters implicate MILI in transposon control. *Science* **316**, 744–747.

2

The Emerging Non-Coding RNA World

Ahmad M. Khalil,[*][†] *Maite Huarte*[*][†] *and John L. Rinn*[*][†][‡]

The post-genomic era has revealed that mammalian genomes produce a wide variety of non-coding RNA transcripts that rival the number of mRNAs produced by known protein-coding genes. In addition to classical RNAs (such as ribosomal RNAs, transfer RNAs and others) and a large class of small non-coding RNAs (such as microRNAs and promoter-associated small RNAs), there is an ever-growing category of large non-coding RNAs. These mysterious RNA molecules share many of the same properties as protein-coding genes: transcribed by RNA Polymerase II, 5′ methylated-Guanine 'cap', contain polyadenylated 3′ end, yet do not code for sensible amino acid sequences. In this chapter we focus on recent advances in understanding large non-coding RNAs. The few well-studied examples have suggested a possible function in numerous biological processes such as embryonic stem cells pluripotency to chromatin guidance and organisation. Also, misregulation of some large non-coding RNAs is observed in many cancers, genetic and neurological disorders, underscoring the importance of these molecules in human health.

*BIDMC Department of Pathology, Harvard Medical School, Boston, MA 02215, USA. †The Broad Institute of Harvard and MIT, Cambridge, MA 02142, USA. ‡E-mail: jrinn@broad.mit.edu.

2.1 Introduction to Non-Coding RNAs

The central dogma of molecular biology which was first introduced by Francis Crick in 1958 has stated that genetic information encoded in DNA is transcribed to form messenger RNA (mRNA) molecules, which then serve as a template to produce unique amino acid sequences that constitute individual proteins (i.e. one gene — mRNA Intermediate — one protein). However, as early as the 1960s, this view began to be challenged by the discovery of non-coding RNAs which do not code for proteins but rather function as RNA molecules (i.e. ribosomal RNA (rRNA) and transfer RNA (tRNA), and small nucleolar RNAs (snoRNAs)). Several new classes and examples of small non-coding RNAs have burgeoned in the post-genomic era as microRNAs (miRNAs) which are involved in translational repression, small interfering RNAs (siRNAs) that regulate RNA transcript levels and Piwi-associated RNAs (piRNAs) which silence transposons during germ cells differentiation. Recent advancements in new technologies such as microarray technology and high-throughput sequencing have revealed a whole new type of non-coding RNA: large ncRNAs (Figures 2.1 and 2.2). Although the first functional large ncRNA, XIST, was discovered in the early 1990s and serves to silence an entire X chromosome in females, it is now clear that there are many more thousands of these large ncRNA molecules encoded throughout the genome. In sharp contrast to the progress made in understanding small RNAs, very little is known about large ncRNAs which consist of intronic ncRNAs, large intergenic non-coding RNAs (lincRNAs), bidirectional ncRNAs and natural antisense transcripts (NATs) which usually overlap with protein-coding genes (see Figure 2.2 and Tables 2.1 and 2.2).

At the beginning of the twenty-first century, new technologies emerged that allowed the examination of RNA transcription on a whole genome level revealing a massive number of large ncRNAs. These RNA molecules are expressed in a wide variety of tissues and cell types, some are developmentally regulated and some become activated in response to specific stimuli. Interestingly, these ncRNA molecules resemble protein-coding genes as they are multi-exonic, spliced, poly-adenylated and 5′ methyl capped, yet do not encode a sensible amino acid sequence. Despite the functional characterisation of a few large ncRNAs, there is

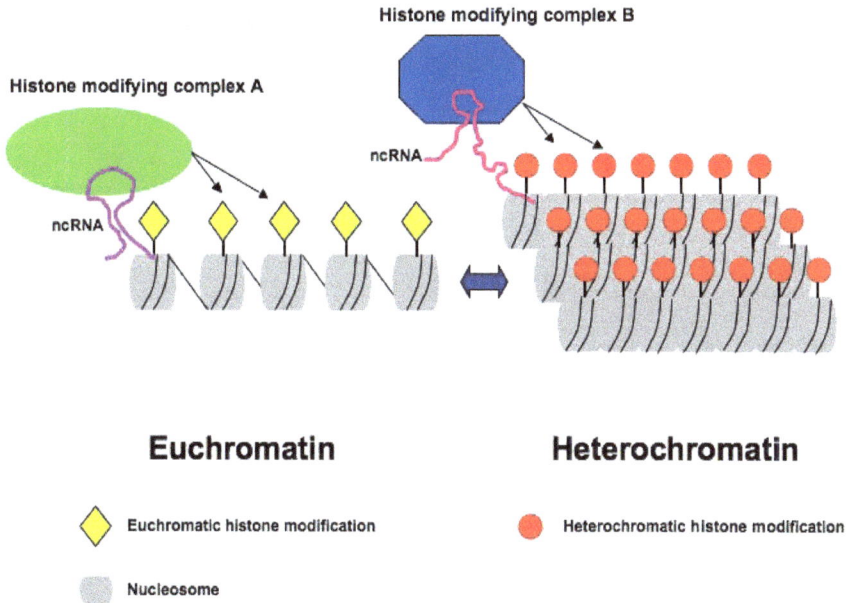

Figure 2.1.   A model of large intergenic non-coding RNAs (lincRNAs) mediated guidance of chromatin-modifying complexes. lincRNAs are transcribed in a similar manner to mRNAs, however, instead of being exported to the cytoplasm to be translated into proteins; many lincRNAs are retained in the nucleus and form ribonucleoprotein complexes with a wide range of proteins including chromatin-modifying complexes. Chromatin-modifying complexes can add covalent modifications to specific residues of histone proteins; however, since the majority of chromatin-modifying complexes do not have DNA binding capacity it has been puzzling how these complexes are targeted to specific genomic loci. In this proposed model, lincRNAs are depicted guiding these complexes to specific genomic loci.

currently a debate in the scientific community as to whether all detected transcripts are functional. There have been three main reasons leading to this scepticism: (a) many of the reported transcripts occur at lower levels than protein-coding genes; (b) some have suggested that these RNA molecules are simply due to spurious transcription. While some of the transcripts show tissue-specific expression, it has been argued that such expression patterns may simply be the result of open chromatin domains;[1] (c) lack of evolutionary conservation. Less than 5% of current ncRNAs catalogues contain RNA molecules with sequence conservation. Although the lack of conservation does not disprove function, it does

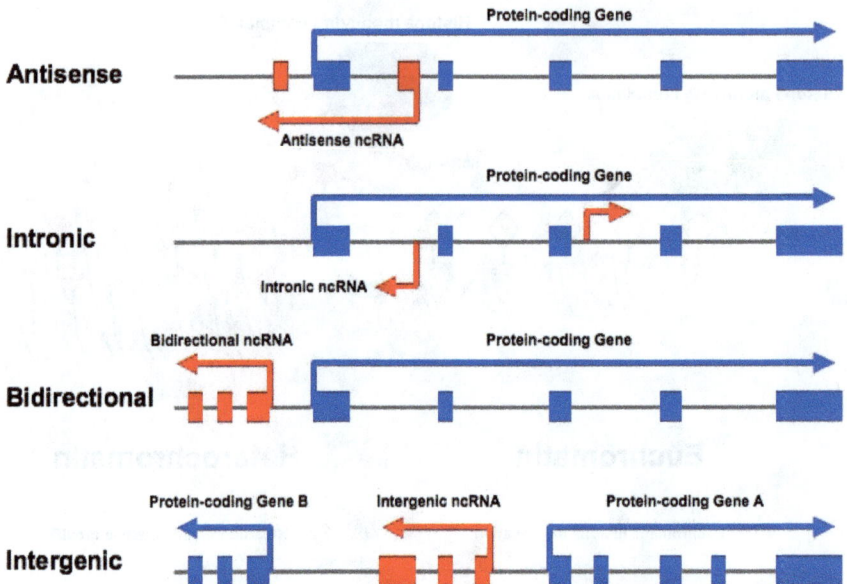

Figure 2.2. **Genomic organization of large non-coding (lnc)RNAs.** lncRNAs are transcribed from various parts of the genome. Some lncRNAs are transcribed antisense to protein-coding genes (NATs: Natural Antisene Transcripts) or from introns of protein-coding genes. Some lncRNAs are transcribed in close proximity to protein-coding genes and in many cases share a promoter with their protein-coding counterparts (bidirectional lncRNAs). Recently, we found several thousand lncRNAs that are transcribed at least 5 kb or more from known annotated protein-coding genes, these lincRNAs (large intergenic non-coding RNAs) are multi-exonic, spliced and polyadenylated similar to mRNAs but do not code for proteins.

Table 2.1.   Genomic organisation of large non-coding RNAs.

	Source	Example	Function
Intronic	Introns of protein-coding genes	NRON	Nuclear import
Antisense	Opposite strand and overlapping with another gene	TSIX	Regulation of XIST in dosage compensation
Bidirectional	Shares a promoter with another gene and transcribed in the opposite direction	FMR4	Anti-apoptotic
Intergenic	In the vast spaces between protein-coding genes	HOTAIR	Chromatin guidance and organisation

Table 2.2.   Small non-coding RNAs in mammals.

	Acronym	Length (nt)	Processed by	Function
MicroRNA	miRNA	18–24	Drosha and Dicer	Regulation of mRNA translation
Small interfering RNA	siRNA	21–23	Dicer	Post-transcriptional regulation of mRNA
Piwi-associated RNA	piRNA	26–34	Unknown	Regulation of transposons in germ cells

put the burden on an experiment to determine if these molecules are truly functional. Nevertheless, there appears to be a consensus that there is a significant amount of RNA transcripts that have not been previously appreciated.

More recently, there has been mounting evidence addressing these three issues and it indicates that a large portion of these RNA molecules are indeed functional:

(a) A large-scale study of large ncRNAs' expression in the brain has shown that these ncRNAs show cell-specific *in vivo* expression patterns within the brain, with some ncRNAs expressed in all brain regions while others are only expressed in specific cell types.[2] The authors also note that 'bidirectional' large ncRNAs do not always show concordant expression with the neighbouring protein-coding genes, and thus could not be a result of spurious transcription. This study strongly suggests that large ncRNAs are not simply due to transcriptional noise from open chromatin domains, but rather points to a potential preponderance of functional large ncRNAs.

(b) It has previously been shown that ncRNAs can be under direct transcriptional regulation of key transcription factors; for example, a recent study found 118 large ncRNAs that are restricted in expression to embryonic stem cells (ES) and 111 of these are directly bound by key ES cells transcription factors such as Sox2, Oct4 and Nanog as determined by independent Chromatin Immunoprecipitation (ChIP) studies. This binding was demonstrated to be functional by cloning the promoter and demonstrating in a reporter assay that this binding is required to activate these ES cells specific large ncRNAs. Thus, numerous large ncRNAs are directly regulated by key transcription factors and are unlikely a reflection of spurious transcription.

(c) A recent study has characterised a class of large intergenic non-coding RNAs (lincRNAs) which are highly conserved and thus likely functional. These lincRNAs appear to have diverse functions although they may share a common mechanism of action (see Section 2.4.1). It is worth noting that some large ncRNAs are not well conserved and yet functional. For example, HAR1 is a RNA gene expressed during cortical development that has evolved rapidly in humans. Thus, the remainder of observed large ncRNAs that are not conserved may be rapidly evolving genes.

Now that a large number of ncRNAs have been identified, it is crucial to begin dissecting the function of these molecules experimentally. In this chapter we will highlight the latest advances in our understanding of the diverse functions of RNA molecules with specific emphasis on large ncRNAs (small RNAs are discussed in subsequent chapters). Dissecting the functional roles of non-coding RNAs is an enormous task but is already leading to findings with notable rewards.

2.2 Classifications of Long ncRNAs Based on Genomic Organisation

The large number of non-coding RNAs that have been sequenced by several consortiums and laboratories around the world have been difficult to characterise under a general theme or class owing to sheer numbers. Moreover, the few long ncRNAs that have been functionally characterised demonstrate a wide range of functional diversity. For example, NRON is a long ncRNA that is important in nuclear trafficking, and AIR and HOTAIR are critical for imprinting loci and HOX gene regulation, respectively. Classifying long ncRNAs has been confounded by only having a handful of functional examples, with diverse biological roles, relative to the plethora of these molecules detected in the genome.

In addition to only having a few functional examples, it has also been problematic to define a large ncRNA primarily due to two problems: (a) ambiguous bioinformatics definitions, and (b) the findings that mRNAs which do make proteins also function as large ncRNAs. Currently, we only have a rudimentary definition of large ncRNAs based on the potential open reading frame length. It has been ambiguously assigned that RNA molecules with any open reading frame greater than 100 amino acids (a.a.) is a 'protein-coding gene'. However, some large ncRNAs are as long as 17–20 kilobases and could have an open reading frame greater than 100 a.a. by chance even if the RNA molecule never produced a protein. Moreover, these open reading frames in large ncRNAs may not produce a sensible a.a. sequence that has been observed across all kingdoms of life. To circumvent this issue it is becoming common practice to use BLASTX[3] to determine if any open reading frames, including those less than 100 a.a., resemble any of the over 240,000 known protein sequences in the Swiss Protein Database. A more recent method has been employed that does not appear to be limited by any of the previously mentioned problems: codon substitution frequency or CSF.[4] The CSF algorithm examines the substitution rate in codons across RNA exons throughout evolution. For protein-coding genes there is clear evidence (high CSF scores) for synonymous codon substitutions resulting in the same a.a. due to codon redundancy, to be highly preserved across the entire coding exon. For known functional large ncRNAs there is a clear signature (negative CSF scores) that evolution is not preserving a.a. content across the exons and therefore they are likely non-coding. Thus, using CSF one can clearly distinguish between protein-coding and non-coding exons.

A second problem in defining large ncRNAs has been the observation that large ncRNAs are encoded within classical messenger RNAs (mRNAs) or that mRNAs act as large ncRNAs under certain cellular contexts. For example, the p53 mRNA can bind to the Mdm2 protein and impair its ubiquitin ligase activity which normally accelerates the degradation of the p53 protein resulting in an increase in p53 abundance.[5] The genes that produce RNA molecules with protein-coding capacity can still function as RNA molecules, further complicating the separation of protein-coding from non-coding genes. Furthermore, some protein-coding mRNA molecules may contain within their 3′ UTR a RNA molecule that functions independently of the protein-coding portion of the molecule. For example, the Oskar gene in Drosophila, which codes for a protein that is required for proper abdomen and germline formation, has a 3′ UTR that functions independently of the Oskar protein during oogenesis.[6]

In light of the difficulty of defining large ncRNAs, they have been typically classified by their genomic loci into four categories: (a) Intronic: encoded within an intron of a protein-coding gene such as NRON; (b) Antisense: transcribed within a protein-coding gene in the opposite direction (e.g. AIR and BACE1-AS); (c) Bidirectional: sharing a promoter with another gene and transcribed in the opposite direction (e.g. FMR3 and FMR4); and (d) Intergenic: those ncRNAs that reside in the vast intergenic spaces of the genome and do not overlap or reside near other genetic elements such as XIST and HOTAIR. In the following section we will discuss each of the four classes of large ncRNAs in great detail and explore the up-to-date functional characterisation of some of these ncRNAs — such as chromatin guidance and formation, regulation of protein-coding genes and extracellular signalling — and how these long ncRNA molecules have been implicated in human disease.

2.2.1 *Intronic non-coding RNAs*

For a long time, large ncRNAs transcribed from intronic regions have received little attention from the scientific community, possibly due to the biased perception that all of them may result from immature mRNAs. But the development of tiling arrays as a tool to detect transcription throughout the whole genome has led to several studies that unveiled many intronic sequences that are independent large and/or small ncRNAs.[3,7] One study demonstrated that 10% of intronic sequences were expressed in human placental tissue, many of which were encoded antisense to the expressed gene. More recent studies have discovered that about 81% and 70% of all spliced human and mouse protein-coding genes, respectively, have transcriptionally active introns, and it has been estimated that more than 30% of un-annotated transcripts reside within intronic regions.[7] Interestingly, one fourth of miRNAs have been found encoded within intronic regions[8] although these only represent a small fraction of the total of intronic ncRNAs.

It has long been disputed whether long intronic transcripts have an independent function from the protein-coding genes transcripts. In many cases, intronic ncRNA-protein-coding mRNA pairs present concordant expression profiles, indicating that at least some of them could be processed from the same pre-mRNA. However, these ncRNAs are stable and most likely functional, so they cannot be considered just artefacts from unprocessed RNA but rather intronic ncRNAs produced from a common protein-coding RNA precursor. In other cases, the expression of the intronic ncRNA might be independent of the protein-coding mRNA. In concordance with this, there is increasing evidence of the presence of independent promoters and transcription factor binding sites within intronic sequences.[7]

Intronic ncRNAs often show distinct expression patterns that suggest their functional relevance is independent of the overlapping protein-coding gene. Analysis of the expression of intronic ncRNAs in several tissues reveals expression signatures that appear to be more tissue-specific than those of well-characterised protein-coding RNAs.[9] Such is the case of the antisense transcript Dnm3os, contained within an intron of the mouse Dnm3 gene. Dnm3os conservation and restricted expression pattern suggests that it may perform an important function during embryonic development.[10] Likewise, *in situ* hyridization has associated intronic ncRNAs with specific neuroanatomical regions, cell types or subcellular compartments of the adult mouse brain.[7] Moreover, the expression of intronic ncRNAs is correlated with different disease states. Some intronic non-coding RNAs correlate to the degree of tumour differentiation in prostate cancer,[11] while others are down-regulated in clear cell renal carcinoma.[12]

There are only a handful of examples of long intronic ncRNAs whose functions have been studied, and similarly to other long ncRNAs these seem to work through a variety of mechanisms. An intriguing example is that of the intronic sequences from the cystic fibrosis transmembrane conductance regulator (CFTR) gene. The overexpression of these three long intronic ncRNAs in HeLa cells causes extensive and specific transcriptional changes, mainly affecting genes linked to CFTR function. Interestingly, each intronic sequence induces unique and highly reproducible changes in gene expression and the affected genes are distributed at spatially diverse sites within the genome.[13] Although their mechanism of function remains unknown, it is tempting to think that similarly to the long intergenic ncRNA HOTAIR,[3] CFTR intronic ncRNAs may affect gene expression through an *in trans* mechanism of epigenetic regulation. Another interesting intronic ncRNA is NRON, found to negatively regulate the nuclear factor of activated T cells (NFAT). It has been proposed that NRON regulates NFAT nuclear trafficking, as it interacts with proteins from the importin-beta superfamily.[14]

It has also been proposed that a number of long intronic ncRNAs are processed into smaller ncRNAs to exert their cellular functions according to previously described mechanisms. This was suggested by recent data revealing that at least one fourth of mammalian miRNA loci is located within intronic regions and is transcribed by RNA Polymerase II. Some snoRNAs are also encoded within intronic regions. Interestingly, a recent work by Zhou *et al.*[15] showed that both human and mouse intronic miRNAs tend to be present in large introns with 5′-biased position distribution, which correlates with the previous observation that most long intronic transcripts are expressed within the first introns of the host genes. The 5′-biased positions of miRNA host introns may be necessary for the transcription and regulation of intronic miRNAs to utilise the regulatory signals within the 5′-UTRs of their host genes.[15]

2.2.2 *Antisense ncRNAs*

Recent studies indicate that a significant fraction of mammalian genes, with estimates ranging from 20% to 72%, show evidence of transcription from both sense and antisense strands.[16,17] These antisense non-coding RNAs are often referred to as Naturally Occurring Antisense Transcripts (NATs). Overlaps of *in cis* sense/antisense transcript pairs can target different portions of the correspondent transcriptional unit, giving rise to three basic types of sense/antisense pairs: head-to-head or divergent; tail-to-tail or convergent; fully overlapping. Although there are some examples of antisense ncRNAs with relevant cellular functions (i.e. XIST/TSIX, AIR, KCNQ1ot1, discussed below), it remains unclear whether most of these natural antisense transcripts have a biological function or if they are just a product of pervasive transcription due to the accessibility of the transcriptional machinery to the open chromatin.

Antisense ncRNAs are capped, spliced and polyadenylated transcripts, potentially forming RNA–RNA hybrids with the processed corresponding sense transcript. Although the effects of antisense transcription may not require the formation of a RNA duplex, different mechanisms by which these transcripts might function have been proposed based on their capacity to form RNA–RNA hybrids with the processed mRNA of the corresponding sense transcript. Some have proposed that they might regulate RNA editing, although experimental evidence argues against this possibility, as RNA editing most likely occurs co-transcriptionally, and antisense ncRNAs often function in their spliced form. Other proposed mechanisms are the interference with transcription or splicing, or the activation of the interferon cascade by binding and activating Protein Kinase R (PKR).[18]

An emerging and attractive model, which has some experimental support, is the role for antisense transcription in RNA interference. In this model, the RNA–RNA hybrids would be processed to yield functional small interfering RNAs (siRNAs) that could serve to silence transposable elements, mRNAs or guide the establishment of heterochromatin. The latter has recently been shown from yeast to mammals. In mammals, random X chromosome inactivation is initiated *in cis* by the long ncRNA XIST, whereas its antisense TSIX blocks silencing. Although the complementarity between XIST and TSIX RNA sequences is not the only determinant of X chromosome inactivation, it has been shown that XIST and TSIX form a RNA hybrid that is processed into small RNAs in a Dicer-dependent manner, and guides the formation of heterochromatin on the inactive X chromosome.[19] The contribution of this mechanism to X-chromosome inactivation remains controversial, as a more recent study using Dicer-deficient embryonic stem cells (ES) argues that Dicer is not essential for initiation of X chromosome silencing.[20] On the other hand, it has been discovered that a 1.6-kilobase ncRNA (RepA) encoded within XIST recruits the Polycomb repressive complex (PRC)2 to the X chromosome and is required for the initiation and spreading of X chromosome inactivation.[21] This particular example underlines the evolutionary conservation of ncRNAs as guidance for heterochromatin formation, a phenomenon that has been well established in plants and fission yeast, where small RNA intermediates facilitate the formation of heterochromatin at centromeres.[22] In summary, it appears that the majority of the, albeit few, known examples of antisense transcripts appear to be involved in chromatin formation. We further discuss the details and mechanisms of some known examples in Section 2.2.2.1.

2.2.2.1 *Antisense ncRNAs in genomic imprinting*

In mammals, several of the known antisense transcripts have established roles in epigenetic silencing phenomena such as mammalian X chromosome inactivation, parental imprinting or monoallelic expression of mammalian immunoglobulins. Here we describe classic examples that have set a precedent for antisense transcription in parental gene imprinting. Although their mechanism of function remains unclear, some evidence points towards the recruitment of chromatin repressive complexes to the imprinted loci by ncRNAs as a potential general theme.

Air: The IGF2 receptor (IGF2R) is an antagonist of IGF2 mitogenic activity. IGF2R exists in a cluster of imprinted genes, and is transcribed only from the maternal chromosome in embryos. Within the second intron of IGF2R resides a promoter that drives transcription from the paternal chromosome of a ncRNA transcript, AIR, which partially overlaps with IGF2R. AIR is essential for the silencing *in cis* of several genes in its locus, including IGF2R. Recently, it has been shown that AIR recruits the H3K9 histone methyltransferase G9a to specifically silence one of the genes (SLC22a3) in the placenta. However, the mechanism of epigenetic silencing for the rest of the locus remains to be understood.[23]

KCNQ1ot1: KCNQ1 is a gene centrally located among eight maternally expressed imprinted genes in a region spanning 1 Mb. Within intron 10 of KCNQ1, resides the promoter for a non-coding, spliced, antisense transcript, KCNQ1ot1. Paternal expression of KCNQ1ot1 induces silencing of the imprinted genes *in cis*. By combining the techniques of Chromatin Immunoprecipitation (ChIP) and RNA Immunoprecipitation (RIP), it has been shown that KCNQ1ot1 interacts with chromatin in regions enriched with H3K9me3 and H3K27me3, and the H3K9- and H3K27-specific histone methyltransferases G9a and PRC2.[24] Like in the case of the AIR–G9a interaction, KCNQ1ot1 interacts with G9a and PRC2 in a lineage-specific manner. It has been proposed that KCNQ1ot1 establishes lineage-specific transcriptional silencing patterns through recruitment of chromatin-modifying complexes and maintenance of these patterns through subsequent cell division occurs via targeting the associated regions to the perinucleolar compartment.[24]

2.2.2.2 Other types of gene regulation by antisense ncRNAs

p15: Expression of antisense ncRNAs is not restricted to imprinted loci, but is a general phenomenon throughout the genome.[25] As the function of diverse ncRNAs is increasingly studied by researchers, the common theme of epigenetic regulation is becoming more evident. The observation that epigenetic alterations are common in cancers has brought the attention to some antisense ncRNAs encoded in tumour suppressor genes. Such is the case of the antisense transcript of the p15 gene, a cyclin-dependent kinase inhibitor implicated in leukaemia. This antisense ncRNA induces p15 silencing both *in cis* and *in trans* through heterochromatin formation. The silencing of p15 by its antisense ncRNA is Dicer-independent, and induces cell growth in embryonic stem cells, suggesting a role in tumourigenesis.[26]

    p21: Another tumour suppressor gene that is subjected to complex regulation by bidirectional (sense and antisense) transcription is p21. It has been proposed that the epigenetic silencing of the p21 gene is regulated by a large non-coding RNA encoded antisense to the p21 mRNA. Morris *et al.* demonstrated that repression of p21 mRNA levels requires both the antisense transcript and Argonaute 1. Specifically, depletion of the antisense transcript or Ago1 results in activation of p21. They propose a model whereby antisense transcription recruits Ago1 to the promoter of p21 and further results in an increase H3K27me3 occupancy thus repressing the transcription of the p21 mRNA.[27]

2.2.3 Bidirectional ncRNAs

In the mammalian genome, there is a prevalence of bidirectional transcription where two transcripts are transcribed in close proximity but in opposite directions.[28] In many instances, one gene is protein-coding while the opposite transcript is a non-coding RNA. In the mouse, approximately 10% of all annotated genes have a bidirectional transcript. These bidirectional transcripts are likely to share promoters and promoter-associated regulatory elements. Also, since these transcripts are usually in close proximity, one would expect that these pairs would show concordant expression. While that is the case for some pairs, other pairs show discordant expression depending on the tissue or the cell type examined. Previously, Mercer *et al.* identified 51 ncRNAs that form bidirectional pairs with protein-coding genes in the mouse brain.[2] Most of these bidirectional pairs show concordant expression patterns which the authors interpreted as meaning that these pairs possibly have similar functions within the brain. For example, the chromatin remodelling gene known as Satb2 and its bidirectional ncRNA are both expressed only in cortical neurons. By contrast, a bidirectional ncRNA associated with Camkk1, a protein-coding gene that functions in male memory formation, exhibits discordant expression in several brain regions.

Both FMR1 and FMR2 are protein-coding genes expressed from the distal end of the long arm of the X chromosome and repression of these genes results in fragile X syndrome and X-linked mental retardation, respectively.[29] Both genes have non-coding transcripts upstream of their transcription start site and in the opposite direction. FMR4 is a 2.2 kb ncRNA transcribed about 200 bp upstream of FMR1 and in the opposite direction.[30] FMR4 and FMR1 show similar expression in some but not all tissues examined. Furthermore, there seems to be no cross regulation between these transcripts since the repression of FMR4 via siRNAs does not influence FMR1 expression and vice versa. However, siRNAs' repression of FMR4, but not FMR1, results in increased apoptosis in two different human cell types suggesting that the ncRNA FMR4 may have an anti-apoptotic function in human cells.[30] FMR3 is also a ncRNA transcribed within several bases from FMR2 and in the opposite direction. However, the function of FMR3 is not known.[31]

Currently, it is not clear if bidirectional transcripts share more than a common genomic locus. It will require further study to discern the significance of the close proximity of these transcripts. Nevertheless, since transcriptional pairs do not always show concordant expression, this provides further evidence that the expression of these ncRNAs is not simply a result of open chromatin structure and that these ncRNA transcripts are likely to be functional.[32,33]

2.2.4 *Large intergenic non-coding RNA (lincRNA)*

While some large ncRNAs are transcribed within or in close proximity to protein-coding genes, some large ncRNAs are encoded in the vast intergenic spaces of the mammalian genome.[3,34,35] Previously, Rinn *et al.* examined gene expression from human HOX clusters (a key family of transcription factors that regulate body patterning) using high resolution tiling arrays and found numerous large intergenic non-coding (linc) RNAs, including a functional lincRNA termed 'HOTAIR'.[3] Interestingly, HOTAIR is transcribed inbetween, but not overlapping with, HOXC11 and HOXC12, yet serves to regulate HOXD genes located on a separate chromosome. Upon depletion of the HOTAIR transcript almost the entire HOXD cluster becomes activated from an otherwise silent state. Moreover, HOTAIR loss of function experiments revealed that Polycomb group complexes were no longer able to bind to the HOXD genes, suggesting that HOTAIR may physically interact with Polycomb to maintain proper regulation of HOXD clusters. In support of this model, HOTAIR was found to co-immunoprecipitate with Polycomb group complex 2 (PRC2). Thus it appears that this interesting form of trans-genomic regulation is accomplished by the lincRNA HOTAIR physically associating with chromatin-modifying complexes and perhaps guiding this complex to the HOXD cluster for proper functioning. Perhaps, this is an emerging theme in large ncRNA biology as XIST and AIR demonstrate similar functions *in cis* (to be discussed later).

Very recently, thousands of additional lincRNAs were discovered by using chromatin marks similar to HOTAIR to identify potential large intergenic non-coding RNAs in several cell types in mice and humans.[34,35] Several studies have previously shown that H3K4me3 marks transcription start sites of genes and H3K36me3 marks transcribed regions by RNA Polymerase II.[36–38] These two marks can reliably identify the promoter and full-length transcript, respectively, of most genes. Using this chromatin signature, thousands of novel large intergenic non-coding RNA (lincRNA) genomic loci were identified.[34,35] These novel lincRNA loci were determined *bona fide* non-coding RNAs by the complete lack of open reading frames and non-synonymous codon substitutions across lincRNA exons. Moreover, hundreds of lincRNAs were confirmed by northern blot and real time PCR (RT–PCR). Around 4,000 highly-conserved large intergenic non-coding RNAs (lincRNAs) were identified using this approach.

One of the most striking features of the lincRNAs identified in these studies was the strong evolutionary conservation of these RNA molecules. In fact, 95% of lincRNAs demonstrate strong evolutionary signatures of conservation. This is a sharp contrast to most large ncRNA catalogues where approximately 5% of the RNA molecules are conserved. Although lincRNAs are not as highly conserved as protein-coding genes, this is likely a reflection of different constraints for RNA structures that relate to their function, whereas protein-coding genes need strong conservation across the entire amino acid sequence as many diseases can occur with a single amino acid mutation (for example, sickle-cell anaemia and Angelman syndrome).[39,40] Thus, despite scepticism that a majority of large ncRNAs are conserved, there does appear to be a class of highly-conserved lincRNAs. This may also indicate that these lincRNAs are not rapidly evolving like other large ncRNAs.

Progress in finding additional examples of functional large ncRNAs has been limited by the lack of methods that predict the function of each RNA, thus hampering these putative functions to be tested by hypothesis-driven experimentation. A novel approach was recently developed to identify lincRNA function using a method that profiled the expression patterns of lincRNAs and protein-coding genes across dozens of tissues and cell types. This allowed the function of each lincRNA to be predicted by the types of protein-coding genes that either correlated or anticorrelated to each lincRNA.[34] This guilt-by-association method demonstrated a wide diversity of cellular functions for lincRNAs such as induction during the innate-immune response, direct regulation by p53 upon DNA damage response and direct regulation by the core embryonic stem cell transcriptional machinery. This study was able to not only predict these functional roles, but also validate hundreds of examples using independent model systems. In Section 2.3 we will discuss the emerging functional themes of large ncRNAs.

2.3 Emerging Functional Themes of Long ncRNAs

Although the exact function of long ncRNAs appears to be diverse, there is not yet an apparent correlation of the function of a long ncRNA and the organisation of its genomic loci (intergenic, antisense, bidirectional or intronic). Thus far the function of any given long ncRNAs has either been shown experimentally or has been extrapolated from specific observations. In Sections 2.3.1–2.3.3 we will discuss the various shown or proposed functions of long ncRNAs as they have emerged over the past few years.

2.3.1 *NcRNAs in chromatin organisation and guidance*

Long non-coding RNAs have provided new insights into a fundamental question in biology: how does the same genome present in every cell encode a multitude of different cellular states that result in a liver cell, heart cell or brain cell (Figure 2.3)? Remarkably, the array of these epigenetic landscapes is established by using ubiquitously-expressed chromatin-modifying complexes. In the post-genomic era it has become well known that there is a code of histone modifications and DNA methylation that represent distinctive epigenetic states.[41,42] Currently there are several chemical modifications that can be added to various residues in the N-termini of histone proteins. Each modification can work with other modifications on the same histone or neighbouring histones to affect the level of chromatin compaction which in turn affects DNA accessibility. However the age-old question still remains: how are the enzymatic complexes placing these marks guided to specific sites under different cellular contexts? It has long been suspected that non-coding RNA molecules guide these complexes to their sites of action.[43]

Indeed, very recent studies have discovered a few examples of large non-coding RNA molecules that guide chromatin formation. Interestingly, all three of these known examples (XIST, HOTAIR and AIR)[3,21,23] confer distinctive epigenetic states, yet share a common mechanism: they physically associate with chromatin-modifying complexes and guide them to specific genomic loci that are crucial for proper cellular function. These examples — XIST, HOTAIR and AIR — have diverse functions in dosage compensation, HOX gene regulation and genomic

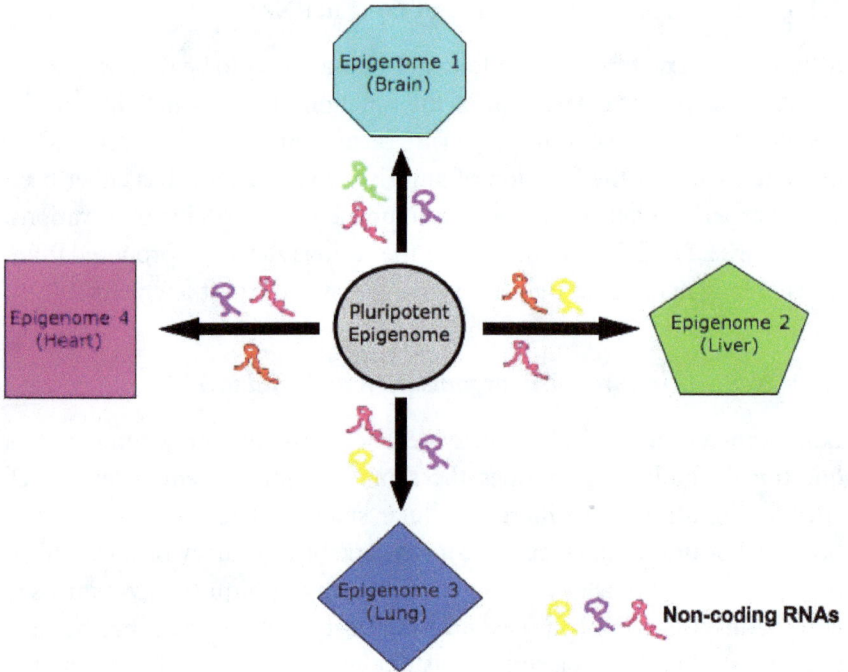

Figure 2.3. Non-coding RNAs could potentially influence cell identity. All cells of multicellular organisms originate from a single totipotent cell, however, the factors that dictate the differentiation pathway of these cells is not well understood. In this model, different types of non-coding RNAs could potentially influence cell identity by influencing gene expression in different cell types. One caveat to this model is how are non-coding RNAs being regulated in the first place.

imprinting, respectively. However, they appear to use a common mechanism: the binding of the long non-coding RNA to a chromatin-modifying complex and guidance to the proper site of genomic regulation.

Mammalian males and females are distinguished genetically by the presence of two copies of the X chromosome in females, and only one copy of the X chromosome in males (in addition to the Y chromosome). Females inactivate one of the two X chromosomes early in development and this remarkable epigenetic phenomenon requires a single large ncRNA. This ncRNA known as XIST (X chromosome inactive transcript) has a length of 19 kb in humans and 17 kb in mice, has no protein coding potential, remains in the nucleus and coats the entire inactive X chromosome *in cis*.[44,45] The exact mechanism by which this non-coding RNA leads to X inactivation (Xi) has remained unresolved for many years. Recently it has been shown that XIST binds to the Polycomb repressive complex (PRC)2 via interaction with the 5′ RepA sequence in XIST with EZH2, the enzymatic component of PRC2. This interaction is required for the proper localisation and silencing of the inactive X chromosome.[19,21] Recently, Ogawa *et al.* have shown that mouse XIST and its antisense non-coding RNA TSIX form RNA–RNA duplexes *in vivo*. The duplex is processed to small RNAs (sRNAs) in a Dicer-dependent manner. Without Dicer, XIST RNA cannot accumulate and histone 3 lysine 27 trimethylation is blocked on the inactive X.[19] By contrast, Kanellopoulou *et al.* have shown that XIST RNA accumulation and H3K27me3 can occur in the absence of Dicer.[20] Further studies are needed to resolve these conflicting results. Nevertheless, collectively these studies suggest that ncRNAs are emerging as guiding molecules for epigenetic regulation of large chromatin domains.

Additional studies have begun to show that non-coding RNAs can potentially be the guiding molecules for these enzymes. Previously, it was demonstrated that HOTAIR, a non-coding RNA in the HOXC locus, can interact with PRC2 and potentially guide PRC2 to the HOCD locus on a separate chromosome.[3] HOTAIR was the first example of a non-coding RNA that can regulate a chromatin domain *in trans*. This finding has major implications in genome regulation where it is possible for a large ncRNA to be transcribed on one chromosome bind PRC2 and ferry this complex around the genome to regulate critical genomic loci. Therefore, misregulation of this process could result in disease. For example, it is well established that loss of HOXD genes results in breast cancer metastasis and HOTAIR is known to be involved in silencing HOXD genes. Thus, ectopic expression of HOTAIR in breast cancer could lead to the inappropriate silencing of HOXD genes resulting in metastasis. The long ncRNA AIR is a third example of a chromatin-associated RNA that guides the establishment of distinctive epigenetic landscapes. AIR is required for the silencing of only the paternal alleles of Igf2r, Slc22a2 and Slc22a3. It was recently demonstrated that this directed silencing is accomplished by the physical association of AIR and the chromatin-modifying protein G9a, a histone

H3K9me2 methyltransferase. This ribonucleic-protein complex is then specifically guided to the paternal Slc22a3, resulting in its silencing.[23]

Collectively, these results suggest that the interface of RNA and chromatin-modifying complexes may be a global mechanism to impart specificity to chromatin-modifying complexes which are ubiquitously expressed and do not appear to have inherent specificity. Indeed a recent study by Khalil *et al.* demonstrated that hundreds of lincRNAs (approximately 20% of expressed lincRNAs in a given cell type) are bound by PRC2 in several human cell types. By examining several other chromatin-modifying complexes, Khalil *et al.* found that 56% of lincRNAs are associated with these complexes. This study further confirmed the localisation of these lincRNAs to chromatin by RNA *in situ* hybridization. Importantly, the study demonstrated, by loss-of-function experiments, that PRC2-associated lincRNAs are required for proper expression of specific regions of the genome which are known to be regulated by PRC2.[35] Thus, this study combined with several previous studies, has revealed a novel RNA-dependent regulation of chromatin states whereby large non-coding RNAs physically associate with chromatin-modifying complexes, conferring epigenetic regulation at specific genomic loci.

Although, this idea of RNA-based chromatin guidance is now becoming clear, it can have a profound effect on our understanding of epigenetic regulation and epigenetic reprogramming during development. For example, the X and Y chromosomes undergo dramatic changes in their histone modifications during spermatogenesis,[46] however the exact mechanism for directing histone-modifying enzymes to these two chromosomes is not known. It is tantalising to postulate that non-coding RNAs mediate such a process. Also, the paternal genome undergoes global DNA demethylation during early embryonic development;[47] the targeting of the paternal genome and the exclusion of the maternal genome are not known. One intriguing possibility is that maternally-inherited non-coding RNAs can protect the maternal genome from this active demethylation and/or guide the active demethylation of the paternal genome. We have only begun to 'scratch the surface' in our endeavour to uncover the role of ncRNAs in epigenetic regulation, but if we could 'crack the code' of RNA-based guidance we could envision engineering RNA molecules that guide these key complexes to genomic loci of our choosing. This would have profound implications in general biology and would also potentially open up new avenues of RNA-based therapeutics.

2.3.2 *Long ncRNAs in differentiation and development*

A number of studies have shown that long ncRNAs: (a) regulate key developmental loci; (b) are temporally and spatially regulated during development; and (c) are important for cellular differentiation. For example, it was recently demonstrated that hundreds of lincRNAs exhibit similar temporal and spatial expression patterns as many key protein-coding genes that are known regulators of development.[34] HOX genes are known to be key regulators of positional identity during development that lead to proper cell fate. It has become clear that HOX genes are regulated by large ncRNAs (i.e. HOTAIR) and that there are hundreds of additional HOX ncRNAs encoded in these loci that exhibit distinctive temporal and spatial expression patterns during development. Another key example is the 6.7 kb large ncRNA TUG1 which is required for the proper development of the photoreceptors in the retina.[48]

Together, the above-mentioned examples have provided a strong precedent for large ncRNAs playing key roles in development, but there is also strong evidence for their regulation of cellular differentiation. For example, in mammals the epithelial to mesenchymal transition is regulated by Zeb-1 and Zeb-2, which in turn regulate E-cadherin expression, a gene essential for maintaining epithelial cells. Interestingly, an antisense transcript regulates the alternative splicing of the Zeb-2 gene that interferes with the translation of the Zeb-2 protein and thus promotes the transition to mesenchymal cells.[49] Thus, cellular differentiation can be mediated by large ncRNAs regulating the translation of key protein-coding genes.

Another role for ncRNAs in cellular differentiation has begun to emerge in the differentiation process of male and female germ cells. In both male and female germ cells, there are a number of large ncRNAs that are exclusively expressed in these cells. Moreover, the Y chromosome has several ncRNAs that are expressed in the male testes.[50] The presence of ncRNAs in male and female germ cells is very interesting since several key epigenetic processes take place in these cells. For example, there is global

DNA demethylation of primordial germ cells (PGC) and re-methylation in maturing germ cells with implications in the epigenetic reprogramming of the genome. Also, it has been proposed that genomic imprints are erased and re-established during germ cell development. The specific expression of long ncRNAs in male and female germ cells suggests that these RNA molecules may contribute to key processes ranging from germ cell differentiation to epigenetic reprogramming. Interestingly, there is also emerging evidence that RNAs can transfer epigenetic information from the gamete to the zygote (paramutation) and can therefore contribute to the regulation of embryogenesis.[50]

Both the mammalian brain and testes have been previously shown to express many of the same protein-coding genes.[51, 52] Another interesting similarity is that both the mammalian brain and testes express high levels of non-coding RNAs. The human brain, for example, contains approximately 100 billion neurons with various morphologies and functions resulting in one of the most intricate networks in a living organism. Previously, Mercer *et al.* examined the expression of long ncRNAs in the mouse brain and found that almost 50% of those analysed are expressed in the brain. Moreover, these ncRNAs show cell-specific gene expression and localisation to specific compartments within neurons.[2] Together, these examples provide mounting evidence that ncRNAs, including long ncRNAs, contribute to important developmental processes ranging from brain development to gamete cell differentiation. We anticipate that future studies will reveal that ncRNAs contribute to a large number of different processes possibly rivalling those of protein-coding genes.

2.3.3 *Promoter associated non-coding RNAs*

An intriguing new model of promoter regulation has begun to emerge recently where large ncRNAs can directly regulate promoter activity. It was previously shown that a large ncRNA is encoded upstream in the promoter region of the Dihydro Folate Reductase (DHFR) gene. This RNA was further demonstrated to directly repress DHFR transcription by forming a stable triple helix with the DHFR promoter and interfering with the transcriptional machinery.[53] Thus, non-coding RNAs encoded 5′ to genes may serve to epigenetically regulate neighbouring genes *in cis*. In support of this model, many recent studies have pointed to the prevalence of 5′ promoter associated RNAs.[54]

A number of studies have shown that many un-annotated transcripts cluster at the 5′ of genes. Most of these transcripts are often smaller than 200 bp and have been called 'promoter-associated sRNAs' (PASRs).[54] In some cases, the transcripts are longer and can include exons of the neighbouring gene: these are named 'promoter-associated long RNAs' (PALRs).[55] Overall, more than 40% of expressed genes have PASR association, while for 80% of silent genes PASRs are not observed.[55] This observation is in concordance with recent work showing that divergent transcription over short distances is common for active polymerase II promoters.[54] Although promoter-associated non-coding RNAs are highly conserved, it may simply be a reflection of promoter sequence conservation. Moreover, it remains unclear what function — if any — they might have. Some have proposed that they are the product of polymerase II abortive initiation, and maintain a chromatin structure permissive to transcription initiation.[56] At present, promoter-associated ncRNAs are new and largely unexplored, but future work on the characterisation of more of these transcripts will likely reveal new paradigms in promoter regulation, such as the observed direct silencing of the DHFR promoter. Collectively, there is mounting evidence that large non-coding RNAs are ripe with functional diversity, possibly rivalling those of protein-coding genes. There is also emerging evidence that perturbations of non-coding RNAs can result in human disease.

2.4 Non-Coding RNAs in Human Disease

As more long, non-coding RNAs are found to be involved in different biological processes, it is not surprising that alterations in those genes are also found related to different pathologies. The availability of new tools such as tiled DNA microarray profiling of non-coding RNAs has started to reveal numerous large ncRNAs misregulated in disease states relative to normal tissue.[57] These and other studies have already implicated numerous large ncRNAs in the aetiology of cancer, neurological disorders and numerous other diseases. We discuss these findings and their implications in Sections 2.4.1–2.4.3.

2.4.1 Long non-coding RNAs in cancer

In recent years, there has been an increasing realisation that cancer cells undergo profound epigenetic alterations with functional consequences in the activity of key genes, and those epigenetic alterations might be as important as genetic defects in the origin of tumours. This realisation has started a new era in cancer research that has changed the vision of the determinants of cancer. One of the breakthroughs in the cancer epigenetics field was the finding that a single small ncRNA, miRNA-10b, was required for breast cancer metastasis.[58] This and other studies[59,60] have provided a precedent for further investigation into ncRNAs in cancer.

Long ncRNAs have also been implicated in cancer. For example, some long ncRNAs are found in ultraconserved regions of the genome that are altered in human leukaemia and carcinomas.[57] Other studies have correlated the expression of intronic non-coding RNAs to the degree of tumour differentiation in prostate cancer or with renal tumours.[11,12] Additionally, a number of NATs have been linked to cancer, like the antisense transcripts of aHIF, EPR-1 of the Bcl-2/IgH hybrid gene.[61-63] Interestingly, several recently discovered lincRNAs are directly regulated by p53 in response to DNA damage response and may be 'tumour-supressor lincRNAs'. Moreover, lincRNAs are functionally associated with oncogenic pathways such as RAS and may represent 'onco-RNAs'.[34] Although there is overwhelming evidence of the involvement of long ncRNAs in cancer, unlike miRNAs, they lack a known common mechanistic pathway, making our understanding of their contribution to disease more difficult. However, more evidence points towards their role as epigenetic regulators and future research in the field will likely explore this exciting avenue.

2.4.2 Long non-coding RNAs in neurological disorders

Non-coding RNAs are overly represented in the central and peripheral nervous system,[64] underscoring the likelihood that nervous system development and function are heavily dependent on RNA regulatory networks, and that perturbations of these networks are the cause of many neurological diseases. Not surprisingly, large ncRNAs have been found to be important to the aetiology of numerous neurological disorders such as spinocerebellar ataxia, Alzheimer's disease, Prader–Willi syndrome, Fragile X syndrome and schizophrenia, all of which are discussed below.

Spinocerebellar ataxia type 8 (SCA8) is a slow form of progressive cerebellar ataxia. There is strong genetic evidence that this neurodegenerative disorder is caused by a triplet expansion within the SCA8 ncRNA.[65] Transgenic mice with the SCA8 expansion mutation develop a progressive neurological phenotype, and Drosophila that over-express human mutant SCA8 show late onset progressive retina neurodegeration.[66] Other long ncRNAs have been implicated in Alzheimer's disease such as BC200, a nervous system specific ncRNA that is involved in the regulation of translation at synapses and subsequent synaptic plasticity.[67] BC200 is up-regulated in the brain of patients diagnosed with Alzheimer's disease, specifically in the regions of the brain associated with this disease.[68] More recently, a ncRNA antisense to the BACE1 gene that regulates the expression of the BACE1 mRNA, has been shown to be elevated in Alzheimer's patients.[69] BACE1 is a critical gene in the brain as it encodes the aspartyl protease required for the initial cleavage of APP to generate amyloid-beta (A-beta).

The Prader–Willi syndrome (PWS) and Angelman syndrome (AS) are two distinct neurodevelopmental disorders, characterised by a range of physiological and neurological anomalies.[70] These syndromes are caused by incorrect imprinting of the PWS/AS region in chromosome 15, in which several ncRNAs are involved, including snoRNAs, miRNAs, IPW, a 2.3 kb long, spliced and polyadenylated non-coding RNA, or the antisense transcript ZNF127 AS.[70] The Fragile X syndrome is the most common cause of inherited mental retardation. It is a genetic disorder that results in a spectrum of physical, intellectual and behavioural features that range from severe to mild in manifestation. The syndrome is associated with the expansion of a trinucleotide (CGG) repeat sequence in the 5′ of the FMR1 gene, and results in the loss of FMR1 gene expression which is required for normal neural function.[71] Recently the involvement of a long ncRNA, FMR4, has been implicated in this syndrome. FMR4 resides upstream of the FMR1 gene and is also silenced in Fragile X syndrome patients. *In vitro* studies have shown that FMR4 affects human cell proliferation and is antiapoptotic, possibly contributing to the Fragile X syndrome by protecting neuronal cells during development.[30]

Additionally, a number of ncRNAs have been associated with schizophrenia, although the only evidence so far is the alteration of these ncRNAs in patients of the disease. Such is the case of PSZA11q14, a ncRNA mapped to chromosome 11q14, a region of chromosomal abnormalities implicated in schizophrenia, which shows reduced expression in brains from schizophrenic patients.[72] An additional example is DISC2 ncRNA, located in the breakpoint region of a chromosomal translocation associated with schizophrenia in a Scottish family.[73] Collectively, these studies provide strong emerging evidence for the roles of long ncRNAs in a wide diversity of neurological disorders. Combined with the observation that over 50% of long ncRNAs are expressed in the brain, these studies give strong precedent for future work to identify other long ncRNAs that are misregulated in human neurological disorders.

2.4.3 *Long non-coding RNAs in other diseases*

Some ncRNAs have also been related to coronary disease. Studies examining single nucleotide polymorphisms (SNPs) associated with disease states have identified a susceptibility locus for myocardial infraction mapped in a long ncRNA, MIAT (Myocardial Infarction Associated Transcript).[74] Likewise, genome-wide association studies identified a region associated with coronary artery disease[75] that includes a long ncRNA, ANRIL (antisense noncoding RNA in the INK4 locus), that associates with a high-risk haplotype for coronary artery disease and which is expressed in tissue and cell types affected by atherosclerosis.

The causes of the epidermal disease psoriasis are unclear, although it is known to have a genetic component. The ncRNA PRINS (psoriasis susceptibility-related RNA gene induced by stress) is a ncRNA transcript that has been associated with this pathology. Over-expression of PRINS is associated with psoriasis susceptibility as PRINS expression is elevated in the uninvolved epidermis of psoriatic patients compared with both psoriatic lesions and healthy epidermis.[76]

Finally, a ncRNA has been related to the human genetic disorder alpha-thalassemia through an epigenetic mechanism of gene silencing. In patients of this disease, expression of a normal alpha-globin gene (HBA2) is silenced as its CpG island becomes completely methylated early during development. It has been shown that transcription of an antisense RNA mediates silencing and methylation of the associated CpG island.[77]

The role of large, non-coding RNAs in human disease is only beginning to emerge. Nevertheless, identifying the role of these non-coding RNAs in the initiation or progression of human disease is an area of great potential in our effort to find treatments and cures for various diseases.

2.5 Conclusion

Only 5% of the human genome is conserved and therefore considered functional. Moreover, protein-coding genes make up only 1.2% of these conserved sequences. Thus, it has long been thought that a majority of the vast intergenic spaces of the genome are mostly comprised of 'junk' DNA. This view of the genome has been drastically changed due to several genome-wide studies using novel technologies such as microarrays that cover most of the genome, and whole genome transcriptomic sequencing technologies which have unexpectedly revealed that a much larger percentage of the genome is indeed transcribed, including intronic and intergenic regions of the genome. While we are still at the initial phase of delineating the functional significance of this large compendium of non-coding RNA molecules, it has become clear that the few that have been studied play major roles in several key processes, including chromatin guidance and organisation, transcriptional regulation, organ development and brain function. Also, it has been observed that alterations in the expression patterns of these ncRNAs can result in several human diseases including cancer and neurological disease. Ongoing research will soon reveal the extent of ncRNA transcription in the genome and will likely yield many more examples of functional RNA molecules in diverse processes, rivalling those of protein-coding genes.

In this chapter we highlighted recent advances in our understanding of long ncRNAs and as current studies point to an emerging role of these non-coding RNAs in chromatin guidance and organisation,[35] perturbation in this fine-tuned system could contribute to human disease such as cancer.

References

1. Struhl K. (2007). Transcriptional noise and the fidelity of initiation by RNA polymerase II. *Nat Struct Mol Biol* **14**, 103–105.
2. Mercer T.R., Dinger M.E., Sunkin S.M. *et al.* (2008). Specific expression of long noncoding RNAs in the mouse brain. *Proc Natl Acad Sci USA* **105**, 716–721.
3. Rinn J.L., Kertesz M., Wang J.K. *et al.* (2007). Functional demarcation of active and silent chromatin domains in human HOX loci by noncoding RNAs. *Cell* **129**, 1311–1123.
4. Lin M.F., Carlson J.W., Crosby M.A. *et al.* (2007). Revisiting the protein-coding gene catalog of Drosophila melanogaster using 12 fly genomes. *Genome Res* **17**, 1823–1836.
5. Candeias M.M., Malbert-Colas L., Powell D.J. *et al.* (2008). P53 mRNA controls p53 activity by managing Mdm2 functions. *Nat Cell Biol* **10**, 1098–1105.
6. Jenny A., Hachet O., Zavorszky P. *et al.* (2006). A translation-independent role of oskar RNA in early Drosophila oogenesis. *Development* **133**, 2827–2833.
7. Louro R., El-Jundi T., Nakaya H.I. *et al.* (2008). Conserved tissue expression signatures of intronic noncoding RNAs transcribed from human and mouse loci. *Genomics* **92**, 18–25.
8. Li S.C., Tang P. and Lin W.C. (2007). Intronic microRNA: Discovery and biological implications. *DNA Cell Biol* **26**, 195–207.
9. Nakaya H.I., Amaral P.P., Louro R. *et al.* (2007). Genome mapping and expression analyses of human intronic noncoding RNAs reveal tissue-specific patterns and enrichment in genes related to regulation of transcription. *Genome Biol* **8**, R43.
10. Loebel D.A., Tsoi B., Wong N. *et al.* (2005). A conserved noncoding intronic transcript at the mouse Dnm3 locus. *Genomics* **85**, 782–799.
11. Reis E.M., Nakaya H.I., Louro R. *et al.* (2004). Antisense intronic non-coding RNA levels correlate to the degree of tumor differentiation in prostate cancer. *Oncogene* **23**, 6684–6692.
12. Brito G.C., Fachel A.A., Vettore A.L. *et al.* (2008). Identification of protein-coding and intronic noncoding RNAs down-regulated in clear cell renal carcinoma. *Mol Carcinog* **47**, 757–767.
13. Hill A.E., Hong J.S., Wen H. *et al.* (2006). Micro-RNA-like effects of complete intronic sequences. *Front Biosci* **11**, 1998–2006.
14. Willingham A.T., Orth A.P., Batalov S. *et al.* (2005). A strategy for probing the function of noncoding RNAs finds a repressor of NFAT. *Science* **309**, 1570–1573.

15. Zhou H. and Lin K. (2008). Excess of microRNAs in large and very 5′ biased introns. *Biochem Biophys Res Commun* **368**, 709–715.

16. He Y., Vogelstein B., Velculescu V.E. *et al.* (2008). The antisense transcriptomes of human cells. *Science* **322**, 1855–1857.

17. Kampa D., Cheng J., Kapranov P. *et al.* (2004). Novel RNAs identified from an in-depth analysis of the transcriptome of human chromosomes 21 and 22. *Genome Res* **14**, 331–342.

18. Werner A., Carlile M. and Swan D. (2009). What do natural antisense transcripts regulate? *RNA Biol* **6**, 43–48.

19. Ogawa Y., Sun B.K. and Lee J.T. (2008). Intersection of the RNA interference and X-inactivation pathways. *Science* **320**, 1336–1341.

20. Kanellopoulou C., Muljo S.A., Dimitrov S.D. *et al.* (2009). X chromosome inactivation in the absence of Dicer. *Proc Natl Acad Sci USA* **106**, 1122–1127.

21. Zhao J., Sun B.K., Erwin J.A. *et al.* (2008). Polycomb proteins targeted by a short repeat RNA to the mouse X chromosome. *Science* **322**, 750–756.

22. Locke S.M. and Martienssen R.A. (2006). Slicing and spreading of heterochromatic silencing by RNA interference. *Cold Spring Harb Symp Quant Biol* **71**, 497–503.

23. Nagano T., Mitchell J.A., Sanz L.A. *et al.* (2008). The Air noncoding RNA epigenetically silences transcription by targeting G9a to chromatin. *Science* **322**, 1717–1720.

24. Pandey R.R., Mondal T., Mohammad F. *et al.* (2008). Kcnq1ot1 antisense noncoding RNA mediates lineage-specific transcriptional silencing through chromatin-level regulation. *Mol Cell* **32**, 232–246.

25. Katayama S., Tomaru Y., Kasukawa T. *et al.* (2005). Antisense transcription in the mammalian transcriptome. *Science* **309**, 1564–1566.

26. Yu W., Gius D., Onyango P. *et al.* (2008). Epigenetic silencing of tumour suppressor gene p15 by its antisense RNA. *Nature* **451**, 202–206.

27. Morris K.V., Santoso S., Turner A.M. *et al.* (2008). Bidirectional transcription directs both transcriptional gene activation and suppression in human cells. *PLoS Genet* **4**, e1000258.

28. Carninci P., Kasukawa T., Katayama S. *et al.* (2005). The transcriptional landscape of the mammalian genome. *Science* **309**, 1559–1563.

29. Hoogeveen A.T., Willemsen R. and Oostra B.A. (2002). Fragile X syndrome, the Fragile X related proteins, and animal models. *Microsc Res Tech* **57**, 148–155.

30. Khalil A.M., Faghihi M.A., Modarresi F. *et al.* (2008). A novel RNA transcript with antiapoptotic function is silenced in Fragile X syndrome. *PLoS ONE* **3**, e1486.

31. Gecz J. (2000). FMR3 is a novel gene associated with FRAXE CpG island and transcriptionally silent in FRAXE full mutations. *J Med Genet* **37**, 782–784.

32. Amaral P.P., Dinger M.E., Mercer T.R. *et al.* (2008). The eukaryotic genome as an RNA machine. *Science* **319**, 1787–1789.

33. Birney E., Stamatoyannopoulos J.A., Dutta A. *et al.* (2007). Identification and analysis of functional elements in 1% of the human genome by the ENCODE pilot project. *Nature* **447**, 799–816.

34. Guttman M., Amit I., Garber M. *et al.* (2009). Chromatin signature reveals over a thousand highly conserved large non-coding RNAs in mammals. *Nature* **458**, 223–227.

35. Khalil A.M., Guttman M., Huarte M. *et al.* (2009). Many human large intergenic noncoding RNAs associate with chromatin-modifying complexes and affect gene expression. *Proc Natl Acad Sci USA* **106**, 11667–11672.

36. Mikkelsen T.S., Ku M., Jaffe D.B. *et al.* (2007). Genome-wide maps of chromatin state in pluripotent and lineage-committed cells. *Nature* **448**, 553–560.

37. Bernstein B.E., Meissner A. and Lander E.S. (2007). The mammalian epigenome. *Cell* **128**, 669–681.

38. Bernstein B.E., Mikkelsen T.S., Xie X. *et al.* (2006). A bivalent chromatin structure marks key developmental genes in embryonic stem cells. *Cell* **125**, 315–326.

39. Coleman E. and Inusa B. (2007). Sickle-cell anemia: Targeting the role of fetal hemoglobin in therapy. *Clin Pediatr (Phila)* **46**, 386–391.

40. Matsuura T., Sutcliffe J.S., Fang P. *et al.* (1997). *De novo* truncating mutations in E6-AP ubiquitin-protein ligase gene (UBE3A) in Angelman syndrome. *Nat Genet* **15**, 74–77.

41. Jenuwein T. and Allis C.D. (2001). Translating the histone code. *Science* **293**, 1074–1080.

42. Strahl B.D. and Allis C.D. (2000). The language of covalent histone modifications. *Nature* **403**, 41–45.

43. Andersen A.A. and Panning B. (2003). Epigenetic gene regulation by noncoding RNAs. *Curr Opin Cell Biol* **15**, 281–289.

44. Avner P. and Heard E. (2001). X chromosome inactivation: Counting, choice and initiation. *Nat Rev Genet* **2**, 59–67.

45. Brown C.J., Hendrich B.D., Rupert J.L. *et al.* (1992). The human XIST gene: Analysis of a 17 kb inactive X-specific RNA that contains conserved repeats and is highly localized within the nucleus. *Cell* **71**, 527–542.

46. Khalil A.M., Boyar F.Z. and Driscoll D.J. (2004). Dynamic histone modifications mark sex chromosome inactivation and reactivation during mammalian spermatogenesis. *Proc Natl Acad Sci USA* **101**, 16583–16587.

47. Mayer W., Niveleau A., Walter J. *et al.* (2000). Demethylation of the zygotic paternal genome. *Nature* **403**, 501–502.
48. Young T.L., Matsuda T. and Cepko C.L. (2005). The noncoding RNA taurine upregulated gene 1 is required for differentiation of the murine retina. *Curr Biol* **15**, 501–512.
49. Beltran M., Puig I., Pena C. *et al.* (2008). A natural antisense transcript regulates Zeb2/Sip1 gene expression during Snail1-induced epithelial-mesenchymal transition. *Genes Dev* **22**, 756–769.
50. Makrinou E., Fox M., Lovett M. *et al.* (2001). TTY2: A multicopy Y-linked gene family. *Genome Res* **11**, 935–945.
51. Rinn J.L., Rozowsky J.S., Laurenzi I.J. *et al.* (2004). Major molecular differences between mammalian sexes are involved in drug metabolism and renal function. *Dev Cell* **6**, 791–800.
52. Rinn J.L. and Snyder M. (2005). Sexual dimorphism in mammalian gene expression. *Trends Genet* **21**, 298–305.
53. Martianov I., Ramadass A., Serra Barros A. *et al.* (2007). Repression of the human dihydrofolate reductase gene by a non-coding interfering transcript. *Nature* **445**, 666–670.
54. Seila A.C., Calabrese J.M., Levine S.S. *et al.* (2008). Divergent transcription from active promoters. *Science* **322**, 1849–1851.
55. Kapranov P., Cheng J., Dike S. *et al.* (2007). RNA maps reveal new RNA classes and a possible function for pervasive transcription. *Science* **316**, 1484–1488.
56. Gilchrist D.A., Nechaev S., Lee C. *et al.* (2008). NELF-mediated stalling of Pol II can enhance gene expression by blocking promoter-proximal nucleosome assembly. *Genes Dev* **22**, 1921–1933.
57. Calin G.A., Liu C.G., Ferracin M. *et al.* (2007). Ultraconserved regions encoding ncRNAs are altered in human leukemias and carcinomas. *Cancer Cell* **12**, 215–229.
58. Ma L., Teruya-Feldstein J. and Weinberg R.A. (2007). Tumour invasion and metastasis initiated by microRNA-10b in breast cancer. *Nature* **449**, 682–688.
59. Cho W.C. (2007). OncomiRs: The discovery and progress of microRNAs in cancers. *Mol Cancer* **6**, 60.
60. Calin G.A., Pekarsky Y. and Croce C.M. (2007). The role of microRNA and other non-coding RNA in the pathogenesis of chronic lymphocytic leukemia. *Best Pract Res Clin Haematol* **20**, 425–437.
61. Yamamoto T., Manome Y., Nakamura M. *et al.* (2002). Downregulation of survivin expression by induction of the effector cell protease receptor-1 reduces tumor growth potential and results in an increased sensitivity to anti-cancer agents in human colon cancer. *Eur J Cancer* **38**, 2316–2324.

62. Cayre A., Rossignol F., Clottes E. *et al.* (2003). aHIF but not HIF-1alpha transcript is a poor prognostic marker in human breast cancer. *Breast Cancer Res* 5, R223–230.

63. Capaccioli S., Quattrone A., Schiavone N. *et al.* (1996). A bcl-2/IgH antisense transcript deregulates bcl-2 gene expression in human follicular lymphoma t(14;18) cell lines. *Oncogene* 13, 105–115.

64. Mehler M.F. and Mattick J.S. (2006). Non-coding RNAs in the nervous system. *J Physiol* 575, 333–341.

65. Moseley M.L., Zu T., Ikeda Y. *et al.* (2006). Bidirectional expression of CUG and CAG expansion transcripts and intranuclear polyglutamine inclusions in spinocerebellar ataxia type 8. *Nat Genet* 38, 758–769.

66. Mutsuddi M., Marshall C.M., Benzow K.A. *et al.* (2004). The spinocerebellar ataxia 8 noncoding RNA causes neurodegeneration and associates with staufen in Drosophila. *Curr Biol* 14, 302–308.

67. Mus E., Hof P.R. and Tiedge H. (2007). Dendritic BC200 RNA in aging and in Alzheimer's disease. *Proc Natl Acad Sci USA* 104, 10679–10684.

68. Lukiw W.J., Handley P., Wong L. *et al.* (1992). BC200 RNA in normal human neocortex, non-Alzheimer dementia (NAD), and senile dementia of the Alzheimer type (AD). *Neurochem Res* 17, 591–597.

69. Faghihi M.A., Modarresi F., Khalil A.M. *et al.* (2008). Expression of a non-coding RNA is elevated in Alzheimer's disease and drives rapid feed-forward regulation of beta-secretase. *Nat Med* 14, 723–730.

70. Horsthemke B. and Wagstaff J. (2008). Mechanisms of imprinting of the Prader–Willi/Angelman region. *Am J Med Genet A* 146A, 2041–2052.

71. Penagarikano O., Mulle J.G. and Warren S.T. (2007). The pathophysiology of Fragile X syndrome. *Annu Rev Genomics Hum Genet* 8 109–129.

72. Polesskaya O.O., Haroutunian V., Davis K.L. *et al.* (2003). Novel putative non-protein-coding RNA gene from 11q14 displays decreased expression in brains of patients with schizophrenia. *J Neurosci Res* 74, 111–122.

73. Millar J.K., Wilson-Annan J.C., Anderson S. *et al.* (2000). Disruption of two novel genes by a translocation co-segregating with schizophrenia. *Hum Mol Genet* 9, 1415–1423.

74. Ishii N., Ozaki K., Sato H. *et al.* (2006). Identification of a novel non-coding RNA, MIAT, that confers risk of myocardial infarction. *J Hum Genet* 51, 1087–1099.

75. Pasmant E., Laurendeau I., Heron D. *et al.* (2007). Characterization of a germ-line deletion, including the entire INK4/ARF locus, in a melanoma-neural system tumor family: Identification of ANRIL, an antisense noncoding RNA whose expression coclusters with ARF. *Cancer Res* 67, 3963–3699.

76. Sonkoly E., Bata-Csorgo Z., Pivarcsi A. *et al.* (2005). Identification and characterization of a novel, psoriasis susceptibility-related noncoding RNA gene, PRINS. *J Biol Chem* **280**, 24159–24167.
77. Tufarelli C., Stanley J.A., Garrick D. *et al.* (2003). Transcription of antisense RNA leading to gene silencing and methylation as a novel cause of human genetic disease. *Nat Genet* **34**, 157–165.

3

MicroRNAs in *C. elegans* Development

Helge Großhans[*,†] *and Almuth E. Müllner*[*]

MicroRNAs (miRNAs) are a novel class of genes that constitute an important layer of gene regulation, directing diverse cellular and developmental processes such as apoptosis, cell differentiation and stem cell maintenance. *Caenorhabditis elegans* has not only been the organism in which microRNAs were first identified, but work in the round-worm has also been fundamental to our current understanding of the biogenesis, molecular and developmental functions of these small RNAs. We will review what is known about the functions of *C. elegans* miRNAs, some of which are highly conserved in both sequence and function in other animals. We will also discuss approaches towards characterising the vast majority of miRNAs lacking assigned functions, particularly the genetic tools and genomic resources that have made *C. elegans* such a powerful system to dissect miRNA biology *in vivo*.

3.1 Introduction

Starting with the initial discovery of microRNAs (miRNAs), *C. elegans* has been an invaluable experimental system to dissect the biology of these

*Friedrich Miescher Institute for Biomedical Research (FMI), Novartis Research Foundation, WRO-1066.1.38, CH-4002 Basel, Switzerland. [†]E-mail: helge.grosshans@fmi.ch.

small regulatory RNAs, including their biogenesis as well as molecular, cellular and developmental functions. In particular, the demonstration that individual *C. elegans* miRNAs prominently regulate diverse cellular processes such as proliferation, differentiation and neuronal function has helped to create an appreciation of their fundamental roles in the regulation of gene expression. We will review what we know about the — surprisingly small — set of *C. elegans* miRNAs with well-characterised functions, some of which appear to be conserved in other animals, including humans. A summary is provided in Table 3.1.

Unexpectedly, a recently completed miRNA deletion project has revealed that individual loss of many of the 154 *C. elegans* miRNAs does not cause overt mutant phenotypes — we will examine potential explanations for this observation. Finally, we will highlight tools and approaches facilitating a more systematic understanding of miRNA regulatory networks and functions, including recently developed 'genomics' resources.

3.2 MicroRNA Biogenesis

In addition to revealing developmental functions of miRNAs, *C. elegans* research has had a major impact on our understanding of miRNA biogenesis and molecular functions *in vivo*, and in the following paragraphs we will briefly review the main issues with an emphasis on contributions made by *C. elegans* research. A more complete account of miRNA biogenesis, which appears largely conserved among animals, is found in Ref. 1.

MicroRNA biogenesis is initiated by transcription of a capped and polyadenylated primary transcript (pri-miRNA) by RNA Polymerase II (Pol II), yielding for instance in the case of *C. elegans let-7* an ~1.7 kb transcript.[2] Consistent with transcription by Pol II, miRNA promoters can also be used to drive messenger RNA expression[3] and are subject to input from diverse transcription factors that might either activate or repress transcription.[4,5]

The RNase Drosha (DRSH-1) together with its co-factor Pasha (PASH-1) cleaves the pri-miRNA to release a stem-loop structure of ca. 70 nt, the pre-miRNA.[6] However, for a subset of miRNAs termed 'mirtrons', found in worms, flies and humans, processing by Drosha is

Table 3.1. Validated miRNA:target pairs in *C. elegans*.

miRNA	Function	Target	Comments	Ref(s).
lin-4	Developmental timing	lin-14	First miRNA:target pair; functions also in lifespan regulation	65, 66, 78
		lin-28	Mammalian homologue regulates let-7 biogenesis	75
let-7	Developmental timing	lin-41	Conserved in vertebrates	53
		hbl-1		88, 89
		daf-12		48
	Vulval development	let-60/ras	Conserved in vertebrates	48
	Vulval development	K10C3.4		46
	Unknown	Others*		48, 90
mir-48/mir-84/mir-241	Vulval development	let-60/ras		100
	Developmental timing	hbl-1		60
	Developmental timing?	daf-12		23
		lin-41		23
lsy-6	Neuronal differentiation	cog-1	Lsy-6 and mir-273 are part of a regulatory loop	106
mir-273	Neuronal differentiation	die-1		107
mir-61	Vulval development	vav-1		115
mir-1	Synaptic function	unc-29	Two subunits of nicotinic acetylcholine receptor (nAChR)	114
		unc-63		
mir-1	Synaptic function	mef-2	Mef-2 is regulated on the post-synaptic site, in the muscle; the effect on neurotransmitter release is indirect	114

* 'Others' include computationally predicted targets that have been partially validated using either reporter gene expression analysis[90] or genetic criteria.[48]

bypassed by the splicing machinery.[7–9] These miRNAs are located in small introns of host genes, with which they are co-transcribed and from which splicing releases the pre-miRNA without any involvement of Drosha. Four mirtrons are currently known in *C. elegans*,[9] none of which has an assigned function.

Irrespective of whether its production depends on Drosha or splicing, the pre-miRNA is subsequently processed by Dicer (DCR-1) to produce a mature miRNA of ~22 nt in length that is incorporated into a miRNA induced silencing complex (miRISC).[10–12] This complex, which contains the Argonaute proteins ALG-1 or ALG-2 at its core,[11,12] then binds to and silences target mRNAs that contain sequences of partial complementarity to the mature miRNA.

Importantly, data mostly from mammalian systems indicate that many or all of the miRNA biogenesis steps can be regulated so that even efficient pri-miRNA transcription need not result in accumulation of a mature, functional miRNA (reviewed in Ref. 13). Although no such mechanisms have yet been revealed in *C. elegans*, they are likely to exist. For instance, mammalian LIN28 and its paralogue LIN28B inhibit both Dicer- and Drosha-mediated processing of miRNAs of the *let-7* family, reducing mature miRNA accumulation.[13] As we will discuss below (Sections 3.6 and 3.7), both *let-7* and LIN28 are conserved in *C. elegans*, where they function in a shared developmental pathway and genetically interact.[14] *C. elegans* LIN-28 might thus also be capable of regulating *let-7* and/or its sequence-related 'sister' miRNAs.

More generally, the temporal expression patterns of mature miRNAs as detected by northern blotting are well reflected in their promoter activities (judged by fusing the putative promoters to green fluorescent protein (GFP)) for some but not all *C. elegans* miRNAs.[15] Moreover, although miRNAs may be co-transcribed in 'clusters', some clustered miRNAs still exhibit distinct accumulation patterns.[15] Thus, as in mammals, at least a subset of *C. elegans* miRNAs is likely to be under substantial post-transcriptional regulation.

Irrespective of the possible conservation of regulatory mechanisms of miRNA biogenesis, biogenesis itself appears to differ in at least two aspects in the worm relative to other animals. Whereas in flies and mammals Exportin-5 mediates nuclear export of the pre-miRNAs to connect nuclear processing of the pri-miRNA by Drosha and cytoplasmic processing of the pre-miRNA by Dicer,[1] there is no Exportin-5 orthologue in *C. elegans*.[16,17] Therefore, it remains to be established how nuclear export of miRNAs is achieved in the worm. Moreover, the *C. elegans* pri-*let-7* is trans-spliced to a short, capped leader sequence, SL1, replacing part of the 5' end of the primary transcript.[2] Trans-splicing also occurs on most *C. elegans* mRNAs, but its functional relevance — for mRNAs as much as miRNAs — is unclear.[18] Trans-splicing might be important for subsequent steps in *let-7* biogenesis, possibly processing by Drosha. In support of this notion, a deletion in pri-*let-7* that extends into the splice acceptor site but leaves the mature miRNA sequence intact prevents accumulation of the mature miRNA.[2] However, it remains to be established if trans-splicing is truly essential for miRNA biogenesis and in particular, if additional pri-miRNAs are trans-spliced.

3.3 The Mechanism of Action of *C. elegans* miRNAs

Mature miRNAs recognise their targets through an antisense mechanism where they bind to partially complementary sequences in their target mRNAs to silence them. However, the actual silencing mechanism has been controversial, and cell-free, cell-based and whole animal experiments have supported diverse, partially incompatible, modes of action.[19] In particular, creating a paradigm for miRNA function, initial work on *C. elegans lin-4* suggested that this miRNA regulated its targets *lin-14* and *lin-28* at a post-transcriptional level by repressing some step after initiation of translation.[20,21] Similar effects were also observed for some human miRNAs and their target reporters; however, substantial evidence from both cell culture experiments and in vitro assays using cell lysates supports two alternative mechanisms of action: block of translation initiation and exonucleolytic mRNA degradation that differs from RNA cleavage by siRNAs (slicing) in that it is not initiated by endonucleolytic cleavage in the miRNA complementary site.[19] These two mechanisms frequently occur together on the same target.

More recent work on *C. elegans* has indeed revealed that *lin-4* and *let-7*, and possibly other miRNAs, repress translation initiation on their target mRNAs while also reducing transcript levels.[22,23] A possible explanation for the discrepancy with the earlier report of stable *lin-4* target mRNA levels[20,21] is provided by the finding that *lin-14* mRNA levels depend on food availability, with elevated levels seen in starved relative to well-fed animals.[24] This effect appears specific for *lin-14* and was not observed with another *lin-4* target, *lin-28*, or the *let-7* target *lin-41*. Intriguingly, although *lin-14* mRNA levels change under starvation conditions, *LIN-14* protein levels do not,[24] suggesting that mRNA degradation might be a secondary effect of silencing, rather than its cause. The primary silencing mechanism might then be repression of translation initiation, followed by transcript degradation, although this remains to be demonstrated.

The precise mechanism of translational silencing and the molecules involved also remain to be identified. The Argonaute proteins ALG-1 and ALG-2,[11,12] which bind the mature miRNAs, as well as the Argonaute-interacting proteins AIN-1 and AIN-2,[23,25,26] homologues of the vertebrate and fly TRNC6/GW182 proteins, are required, but how they achieve silencing is not well understood. Moreover, although the translation initiation factor eIF6, which prevents premature assembly of the 60S and 40S ribosomal subunits, has been implicated in miRNA function in *C. elegans* and human HeLa cells,[27] other studies in mice, *D. melanogaster* and *C. elegans* have indicated that eIF6 might not be generally required for miRNA function.[28–31] It has been speculated that the involvement of eIF6 may be indirect, possibly reflecting a role in 60S subunit biogenesis.[19] Finally, VIG-1 and TSN-1 are two additional miRISC-associated factors[32] which have also been found to assemble on mature *let-7* in a *C. elegans* lysate *in vitro*, along with other proteins of unknown identity.[33] Whether and how these proteins contribute to miRNA activity is, however, not known, and this is also true for a large number of genes that have been reported to affect miRNA activity based on genetic read-outs.[31,34,35]

3.4 MicroRNA Targets and Their Prediction

Regardless of the precise mechanism that miRNAs employ to repress their targets, antisense binding to mRNA sequences of partial complementarity is required. In fact, miRNAs themselves might primarily function as guides that recruit the RISC complex to target mRNAs, as evidenced by the observation that, in cultured cells, tethering Argonaute or their interacting

GW182 proteins artificially to a mRNA induces silencing of this mRNA.[36,37] However, mutations in the *let-7* target sites of the *C. elegans lin-41* 3′ UTR that were predicted to stabilise this miRNA:target duplex, and thus improve Argonaute recruiting, abolish target gene regulation.[38] It thus appears likely that additional levels of control of miRNA function exist so that, *in vivo*, Argonaute/RISC recruitment may not be sufficient for function. This might explain why computational procedures for miRNA target gene prediction have been frustratingly inefficient, particularly in *C. elegans*.

Although a general agreement exists that the miRNA seed (i.e. the six to eight 5′ most nucleotides) is particularly important for miRNA target gene recognition and/or silencing, such seed matches are not sufficient for reliable target prediction.[39,40] For instance, when 13 predicted targets of the *lsy-6* miRNA with phylogenetically conserved seed match sites were tested for repression by endogenous *lsy-6*, none appeared to be regulated.[40] On the other hand, physiologically relevant target sites might not have perfect seed matches, as demonstrated by the *let-7* target *lin-41* or the *lin-4* target *lin-14*.[41,42] The detailed dissection of a number of miRNA:target interactions in *C. elegans* suggest that the local structure of the 3′ UTR sequence surrounding the target site, but possibly also specific sequence elements help to determine whether specific mRNAs may be regulated by partially complementary miRNAs or not.[42–45] Thus, a 27 nucleotide sequence between two *let-7* complementary sites is required for repression of *lin-41* by *let-7*.[42] Similarly, although two *lsy-6* complementary sites in *cog-1* mediate repression by *lsy-6*, transplanting one or both of them into a non-regulated 3′ UTR is insufficient to confer regulation.[40,43] Indeed, additional ill-defined contextual features in the *cog-1* 3′ UTR are required for *lsy-6*-mediated gene silencing.[43] How these sequences contribute is unknown, although the binding of co-factors — or antagonists — of miRNAs is a possibility.

Experimental strategies have been used to enrich putative miRNA targets before submitting these to sequence analysis for computational target prediction. For instance, miRNAs may be used to prime reverse transcription on their target mRNAs, followed by cloning and sequencing of the amplification product.[46] The beauty of this approach is that it not only has the potential of identifying targets of specific miRNAs, but that it can even provide information on the target site used. However, only targets with paired miRNA 3′ ends can be discovered, which might explain the fact that although a novel *let-7* miRNA target was discovered, none of the known targets were actually recovered. In another approach, miRISC is precipitated, using AIN-1 and AIN-2 immunoprecipitation, and co-precipitating mRNAs identified by microarray analysis.[26] This approach cannot determine the identity of the 'cognate' miRNAs that regulate the enriched mRNAs, but it reduces the sample complexity (miRISC-enriched mRNAs versus all mRNAs) to improve subsequent computational target prediction. The resulting tool named 'miRNA targets by weighting immunoprecipitation-enriched parameters'[47] performs well when benchmarked against known miRNA targets, although no validation of *de novo* identified targets has yet been reported. Finally, a 'classic' approach towards miRNA target identification in *C. elegans* involves the use of genetics to either validate a predicted miRNA target by establishing genetic interaction between the miRNA and its target,[48] or conversely, using unbiased genetic searches for miRNA interaction partners to enrich for potential miRNA target genes,[31] which can then be analysed computationally.

An important strength of genetic suppression for miRNA target identification is that it not only provides a first piece of validation of such a miRNA:target relationship, but also hints at the physiological importance of the identified targets. This is important because even an algorithm that could reliably predict every miRNA target would not be able to provide evidence for the physiological relevance of such regulation. In other words, it is possible that an mRNA is regulated by an miRNA, but that this regulation is inconsequential because residual target gene activity is sufficient to execute the specified cellular function, making this gene a 'neutral target'.[49] However, if one can demonstrate that overexpression of an individual miRNA target — to levels close to those seen upon loss of the miRNA — phenocopies aspects of the miRNA loss-of-function (*lf*) phenotype and, conversely, that mutation or depletion of this target rescues aspects of the miRNA mutant phenotype, physiological (or at least pathological) relevance is established.

The original insight that miRNAs function as antisense regulators of mRNAs was in fact guided by this type of genetic interaction.[50–52] The identification of *lin-14* as a target of the *lin-4* miRNA was based on the genetic interaction and mutant phenotypes of these two genes, in particular the fact that *lin-14* gain-of-function (*gf*) alleles phenocopied, whereas *lin-14(lf)* alleles caused the 'opposite' phenotype of loss of the *lin-4* miRNA (Section 3.6). Subsequently, the *lin-41* mRNA was similarly identified as a target of the *let-7* miRNA through genetic suppression of the *let-7* mutant phenotypes.[53,54]

A caveat for genetic suppressor screens to work is that a phenotype caused by loss of the miRNA of interest is needed, which may not always be the case (Section 3.11). In Section 3.5 we will discuss those miRNAs for which developmental or cellular functions have been elucidated.

3.5 'Small Temporal RNAs' and the Heterochronic Pathway

Proper timing of developmental events is fundamental to normal development, and in *C. elegans* temporal patterning is under explicit genetic control.[55,56] The heterochronic pathway constitutes a molecular timing mechanism that controls the proper execution of temporal cell fates during postembryonic development, which proceeds through four larval stages, L1 through L4, followed by the adult stage. Thus, precocious heterochronic mutations cause cells to skip particular cell fates to adopt fates that would only occur later during wild-type animal development. Conversely, retarded heterochronic mutations cause the reiteration of developmental cell fates relative to wild-type animals. As *C. elegans* development proceeds along an invariant cell lineage with particular cell fates occurring at the same place in time and space in every animal,[57,58] heterochronic mutations result in readily recognisable phenotypes.[59] For instance, the seam cells (a subset of skin cells) divide a specified number of times during the larval stages, before exiting the cell cycle at the larval-to-adult transition, when they also fuse to form a syncytium (Figure 3.1). The syncytial seam cells contribute to the formation of the adult alae, a characteristic cuticular structure the presence of which is frequently used as a proxy for terminal seam cell differentiation. This set of events is termed the larval-to-adult (L/A) switch,[51] and failure to execute this switch can thus be recognised through the occurrence of extra seam cell divisions, and delay or lack of formation of the seam cell

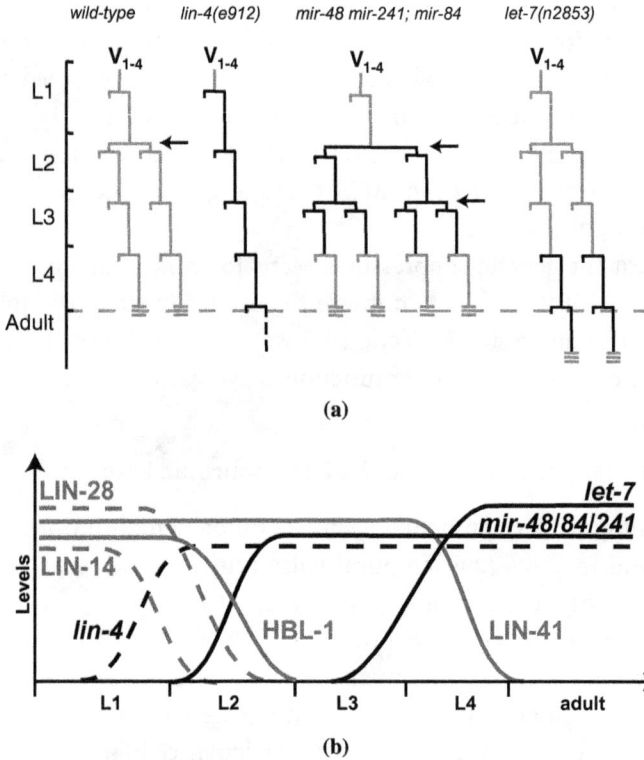

Figure 3.1 (a) Cell lineage of the V1 through V4 seam cells in wild-type and heterochronic mutant animals. Following asymmetric division, anterior (left) cells differentiate and fuse with the *hyp7* syncytium, whereas posterior cells maintain seam cell fates. During the L2 stage, seam cells undergo one symmetric (proliferative) division (*arrow*), increasing the seam cell number. Failure to execute L2 fates (e.g. in *lin-4* mutant animals), or repeated execution of L2 fates (in *mir-48 mir-241(nDf51)*; *mir-84(n4037)* triple mutant animals) will alter seam cell numbers permanently. At the L4-to-adult-transition, seam cells exit the cell cycle, fuse into a syncytium, and differentiate in wild-type animals (*three horizontal bars*), but fail to do so in *let-7(n2853)* animals, leading to a transient increase in seam cell number. Reiterated cell fates are shown in black, wild-type cell fates in grey. Note that the *lin-4(e912)* mutation causes an indefinite reiteration of L1 cell fates, whereas the *mir-48 mir-241(nDf51)*; *mir-84(n4037)* mutation causes only a single repetition of L2 fates. Temperature-sensitive *let-7(n2853)* mutant grown at semi-permissive temperature (15°C; these animals will die at a higher temperature) similarly repeat L4 fates only once, possibly due to residual *let-7* activity. (b) Schematic depiction of expression patterns of selected heterochronic miRNAs (*black*) and their targets (*grey*). *Dashed* lines represent *lin-4* and its targets, *solid* lines miRNAs of the *let-7* family and selected targets. Additional targets are omitted for clarity; cf. Table 3.1.

syncytium and/or the alae. In contrast, precocious phenotypes may cause a reduced number of seam cells relative to wild-type animals, because cell division events are skipped. Cell cycle exit, syncytium and/or alae formation may also occur prematurely. These events can be observed directly using Nomarski microscopy to follow individual nuclei or through the use of appropriate GFP markers.

As discussed in more detail below (Sections 3.6 and 3.7) and elsewhere in this book (Chapter 1), *lin-4* and *let-7* were both discovered through genetic screens for heterochronic genes.[54,59] With additional heterochronic miRNAs now discovered,[60,61] a very simplified model of the heterochronic pathway would invoke a miRNA-target cascade where key protein regulators act to specify events at a particular stage. Expression of specific miRNAs at various times then eliminates these proteins to permit transition to the 'next' fate. However, as we will discuss next, *lin-4* and *let-7* each regulate several targets, and the same heterochronic genes appear to receive regulatory inputs from different miRNAs at different times in development, either in the same or different tissues. Thus, rather than establishing a regulatory cascade, the miRNAs of the heterochronic pathway appear to be part of an extensive regulatory network that involves coordinated regulation of expression of multiple targets across different tissues, by various miRNAs.

3.6 The *lin-4* miRNA Family

3.6.1 *Lin-4 promotes L2 fates by repressing lin-14*

The *lin-4* miRNA was identified in a screen for mutations that caused temporal alterations in postembryonic cell lineages of *C. elegans*, i.e. heterochronic phenotypes (see Section 3.5).[59] Specifically, in *lin-4(lf)* animals, L1 cell fates were reiterated in subsequent stages. For instance, the V1 through V4 seam cells undergo one asymmetric division in the L1 stage, yielding one seam cell and one cell that will fuse with *hyp7* and lose seam cell identity (Figure 3.1). In the L2 stage, on the other hand, these seam cells will first undergo a symmetric division, yielding two seam cell daughters, followed by an asymmetric division. In the absence of *lin-4*, V1 through V4 fail to undergo the symmetric division during the L2, or indeed any subsequent stage.[59] As a consequence, *lin-4(lf)* animals fail to execute the L/A switch. Analogous aberrations in temporal fates are seen in other tissues.

Loss of function of another heterochronic gene identified in these screens, *lin-14*, caused the opposite phenotype, i.e. skipping of L1 fate and premature execution of L2 cell fates, and precocious execution of the L/A switch at the L3 molt.[51,62,63] Moreover, *lin-14(lf)* mutations suppressed loss of *lin-4*, placing *lin-14* downstream of *lin-4* by genetic criteria.[51] Finally, *lin-14(gf)* and *lin-4(lf)* alleles resulted in largely similar retarded heterochronic phenotypes, leading to the conclusion that *lin-4* negatively regulated *lin-14* (Refs. 51 and 52) — a finding of extraordinary importance that was the starting point for our understanding of miRNAs as negative regulators of gene expression.

The identification of *lin-14(gf)* alleles in these screens proved particularly fortunate, as the responsible mutations were found to delete parts of the *lin-14* 3′ UTR.[50,64] Together with reporter gene experiments demonstrating that the *lin-14* 3′ UTR can confer *lin-4*-dependent, post-transcriptional gene silencing and the realisation of partial complementarity between the *lin-4* miRNA and multiple sites in the *lin-14* 3′ UTR,[65,66] the concept of antisense regulation of mRNAs by small RNAs was born.

Consistent with its function in regulating the transition from L1 to L2 cell fates, *lin-4* begins to accumulate after mid-L1 stage and high expression levels are seen from late L1 onwards throughout larval development (Figure 3.1b).[67] Conversely, LIN-14 protein levels are high in the embryo and early L1 larva, but reduced at L2 (Refs. 20 and 68). Overexpressing *lin-4* from a transgenic array causes precocious downregulation of *lin-14* expression and precocious phenotypes in the seam cells and the vulval precursor cells[67] although, unexpectedly, overexpression of *lin-4* following deletion of its transcriptional repressors does not (see below).[4] LIN-14 therefore acts during early L1 to promote L1 fates, and its repression by *lin-4* permits subsequent progression into L2.[63] Low levels of *lin-14* activity are further required during late L1 to specify L2 fates and to prevent premature progression to L3 fates[63,69]; therefore, it has been speculated that LIN-14 might act as a morphogen causing different fates from cells depending on its concentration.[69]

LIN-14 is a transcription factor that represses transcription of the insulin-like gene *ins-33* (and presumably other genes) during the L1 stage by binding to the *ins-33* promoter.[68,70,71] Repression of *lin-14* by *lin-4* permits *ins-33* expression during L2 (Ref. 71). However, *ins-33* is not known to function in temporal patterning, so that the key targets of LIN-14 for temporal specification of cell fates remain to be identified.

As discussed above (Section 3.2), miRNA biogenesis is a complex process that includes several processing steps, all of which can be regulated.[13] A *lin-4::gfp* transcriptional reporter containing only DNA sequences upstream of the miRNA, that is, the putative promoter, approximates temporal expression seen for the mature *lin-4* RNA by northern blot analysis,[4,72] suggesting predominantly transcriptional regulation. However, whereas mature *lin-4* RNA accumulates by late L1, GFP in the seam cells was first observed at the L2 stage.[72] Thus, even though the promoter activity of this short (~515 bp) stretch of DNA is sufficient to rescue *lin-4(lf)* animals when used to drive *lin-4* RNA expression, additional regulatory elements might still be absent.

Recently, Ow *et al.*[4] identified three transcription factors of the FLYWCH family, FLH-1, FLH-2 and FLH-3, as redundant repressors of *lin-4* expression during *C. elegans* embryogenesis. These transcription factors are highly expressed in embryos but downregulated soon after hatching, an expression pattern inverse to that of *lin-4*. By yeast 1-hybrid analysis, FLH-1 can bind to a *lin-4* promoter fragment that is important for correct *lin-4* expression and that carries a putative FLH-1 consensus-binding site, as well as to sequences outside the consensus site. Although deregulation of *lin-4* is particularly strong in various *flh-1*, *flh-2* and *flh-3* mutant backgrounds and combinations thereof, additional miRNAs, including the *let-7* sisters *mir-48* and *mir-241* (Section 3.7) are upregulated in these strains, whereas other miRNAs are downregulated.[4] Indeed, combined loss of FLH-1 and FLH-2 causes lethality, which cannot be suppressed by loss of *lin-4*, combined loss of *mir-48* and *mir-241*, or even *alg-1*, which would be expected to impair the activity of all miRNAs, indicating additional functions of the FLH proteins, independent of miRNA regulation. Possibly reflecting such additional function, elevated *lin-4* levels in FLH mutant embryos and larvae are not sufficient to confer postembryonic heterochronic defects nor a significant reduction of larval LIN-14 levels,[4] although *lin-4* overexpression from a transgenic array was previously reported to cause precocious heterochronic phenotypes.[67] However, it is also possible that different levels of *lin-4* overexpression resulted from these two strategies.

Interestingly, *lin-4* expression and activity are responsive to food availability, although neither the mechanisms nor consequences of this observation are understood in much detail. *Lin-4* promoter activity and mature miRNA accumulation can be repressed by starvation (A. Esquela-Kerscher, personal communication; Refs. 67 and 73). Moreover, adverse environmental conditions such as starvation induce development of L2 stage larvae into developmentally arrested dauer larvae, which can resume development as L4 larvae when conditions improve. Unlike *lin-4* mutant animals that have developed continuously and fail to execute the L/A switch in the seam cells, post-dauer *lin-4(lf)* animals execute this switch normally.[74] Similarly, whereas the L/A switch occurs prematurely in continuously developing *lin-14(lf)* animals, development through the dauer stage suppresses this phenotype.[74] Finally, under certain conditions, starvation appears to leave mature *lin-4* levels unaffected, but alters the levels of the *lin-14* RNA, but not LIN-14 protein.[24] Unravelling these connections between *lin-4*, and the heterochronic pathway more generally, and the environment might be a worthwhile undertaking.

3.6.2 *Lin-4 represses lin-28 to promote L3 fates*

A second target of *lin-4* is *lin-28* (Refs. 51 and 75), a predominantly cytoplasmic protein containing a cold-shock domain and two CCHC-type zinc finger RNA binding motifs. Intriguingly, mammalian LIN28 (and its paralogue LIN28B) have been shown to regulate the expression of *let-7* family miRNAs by controlling their processing (reviewed in Ref. 13). Although not yet shown, it thus appears possible that *C. elegans* LIN-28 might play a similar role, directly linking the early, *lin-4*-dependent, and the late, *let-7*-dependent, timer.

Similar to *lin-14(lf)* animals, *lin-28(lf)* animals display precocious heterochronic phenotypes; however, in this case L2 rather than L1 cell fates are skipped.[62] This observation is consistent with the fact that expression of *lin-28* persists through the mid-/late-L2 stage, when LIN-14 protein has already become largely undetectable.[20,21] Although the causes of this interesting observation remain to be elucidated, it might be a consequence of the smaller number of *lin-4* 3' UTR target sites in *lin-28* (one site)[75] compared to *lin-14* (seven sites),[41] which might require different amounts of *lin-4* for efficient target repression.

Lin-28 and *lin-14* also promote each other's expression, possibly indirectly by repressing repressors of each other's expression.[21,52] Both *lin-14* and *lin-28* are potential targets of the *let-7* family of miRNAs,[48,53,76] although further validation is needed.[23,60] Thus, an attractive, but speculative possibility is that this regulatory circuit involves repression by *let-7* family members. Indeed, mammalian LIN-28 is regulated by both *let-7* and the *lin-4* homologue *miR-125* (Ref. 77).

3.6.3 *Regulation of ageing by lin-4*

In addition to its function in temporal patterning, *lin-4* also regulates ageing, with *lin-4(lf)* mutant worms having a decreased lifespan, and *lin-4* overexpressing worms an increased lifespan, relative to wild-type animals.[78] This function of *lin-4* is also mediated by *lin-14*, as *lin-14(lf)* mutants show the opposite phenotype of *lin-4(lf)* mutants and can suppress the *lin-4* phenotype, whereas *lin-14(gf)* alleles cause a short lifespan phenotype, like *lin-4(lf)* alleles. Lifespan extension is observed even if *lin-14* is depleted by RNAi only in adult animals, pointing towards a novel function of LIN-14 during the adult stage.[78] As previous work found LIN-14 protein to be undetectable by western blotting by the L3 stage (e.g. Ref. 20), this finding is rather unexpected but may simply reflect limited sensitivity of detection of LIN-14 in whole animal lysates when expression may be restricted to a subset of tissues. Indeed, weak *lin-14* expression in some head neurons was observed as late as the L3 or L4 stage with a functional LIN-14/GFP fusion construct.[70] However, it has not yet been demonstrated that it is in these cells that *lin-4* represses *lin-14* to regulate lifespan and/or whether *lin-4* levels are differently controlled in these tissues from, for example, hypodermis.

The insulin-/insulin-like growth factor-signalling pathway is a major regulator of lifespan and genetic evidence supports the notion that *lin-4* affects ageing through this signalling pathway.[78] In this light, regulation of the insulin-like gene *ins-33* by LIN-14 (Section 3.6.1; Ref. 71) gains a new significance, although a role in ageing remains to be demonstrated.

3.6.4 Members of the lin-4 family in C. elegans and beyond

Lin-4 is similar in sequence to mir-237, the only other known member of the C. elegans lin-4 family.[79] However, little is known about mir-237 beyond its expression pattern, which is also temporally regulated during larval development, with the mature miRNA mir-237 first detected at low levels during L2, increasing at early L3, and peaking during the L4 stage.[72] Consistent with this pattern of accumulation, robust mir-237 promoter activity is first observed during early L3, e.g. in the seam cells.[72] Nonetheless, promoter activity was noted as early as the L1 stage in the Z1 and Z4 cells of the gonad. Such restricted expression would be expected to be below the level of detection by northern blotting, although it remains equally possible that the putative miRNA promoter fused to GFP lacked negative regulatory elements that prevent endogenous mir-237 expression in these cells and/or at this time.

In addition to their similar temporal expression profiles, lin-4 and mir-237 promoters exhibited partially overlapping spatial expression patterns, suggesting the possibility of redundancy of these two genes. However, as mir-237 target genes have not yet been validated, it remains to be seen to what extent the two miRNAs regulate an identical or divergent set of target genes in various tissues. Unlike for lin-4, no overt phenotypes were noted in a mir-237 mutant strain examined in a large-scale miRNA deletion screening project,[80] although it remains possible that more specific phenotypes were missed or that background mutations in this strain might have modified the phenotype (Section 3.11).

The identification of lin-4 as the first miRNA[65] and the realisation that miRNA-mediated regulation of gene expression occurs abundantly in animals[81–83] was separated by a lag phase of several years, at least in part owing to the fact that no homologue of lin-4 was initially detectable in any animal species other than nematodes. Ironically, however, subsequent miRNA cloning experiments identified mammalian miR-125a and miR-125b-1/miR-125b-2 as two miRNAs similar in sequence to lin-4 (Ref. 84) that also appear to be functionally conserved. In vitro data have confirmed mammalian LIN28 to be regulated by miR-125 and, consistent with the function of lin-4 in regulating cell proliferation and differentiation, miR-125 expression is initiated upon differentiation of embryonic stem cells and embryonal carcinoma cells.[77,85] With its target LIN28 known to be a pluripotency inducing factor,[86] it thus appears that miR-125, like *lin-4*, promotes cell differentiation.

3.7 The *let-7* miRNA Family

3.7.1 *Let-7 regulates several targets in multiple tissues to promote adult cell fates*

Although *lin-4* was the first miRNA to be discovered, *let-7* subsequently became the poster-child of the field. In contrast to *lin-4*, *let-7* orthologues were readily detected in a wide range of species, exhibiting a perfect conservation of the mature miRNA sequence.[87] This provided the first piece of evidence that miRNA-mediated gene silencing was not limited to *C. elegans* but was instead a major phenomenon in animal development, and *let-7* has subsequently become a widely-used model miRNA to investigate nearly every aspect of miRNA biology. Moreover, it subsequently turned out that the dramatic developmental defects seen in worms lacking functional *let-7* foreshadowed the importance of mammalian *let-7* as a tumour suppressor gene and regulator of stem cell fates.[14]

Similar to *lin-4*, *let-7* was found in a screen for heterochronic mutants.[54] *Let-7* miRNA is required for the larval-to-adult transition, at which major changes in various tissues occur in a coordinated fashion. *Let-7(lf)* mutants die by bursting through the vulva at this time point, apparently due to the combined upregulation of several *let-7* target genes. Viability of mutant animals can be restored when target gene function is impaired through *lf* mutations or depletion by RNAi. Surprisingly, RNAi-mediated depletion of any one of a dozen known *let-7* target genes can partially suppress the *let-7* lethality phenotype,[48,53,54,88,89] although additional targets appear to exist whose depletion does not restore viability.[90]

The apparently indiscriminate suppression of *let-7* by depletion of various targets might be explained by extensive redundancy of target genes, although the fact that target genes act in at least three different tissues, i.e. the ventral nerve cord, the intestine and the seam cells, argues against this possibility.[48,53,88,89] More probably, it is only the combined overexpression of many target genes that makes the *let-7* mutation lethal and not the overexpression of individual genes. Consistent with a threshold model,[91] decreasing the function of any one of these genes might thus suffice for survival.

Arguably the best-characterised target of *let-7* is lin-41, an RBCC-NHL (Ring finger, B-box, Coiled-Coil; NCL-1, HT2A, LIN-41) domain protein of unknown function.[53] As was the case for the first targets of the *lin-4* miRNA, *lin-41* was identified as a target of *let-7* by virtue of its suppression of *let-7* mutant phenotypes, including vulval bursting and retarded seam cell differentiation, and *lin-41*(lf) mutant animals display the opposite, precocious seam cell differentiation phenotype.[53,54] Although the *lin-41* 3′ UTR contains several putative *let-7* complementary sites, only two of these sites appear essential for function *in vivo*.[42] Strikingly, regulation of *lin-41* (known as 'Trim71' in vertebrates) appears to be conserved throughout the animal kingdom, in zebrafish, mice, chicken and humans.[77,92–97] Although the precise functions of Trim71 are not known, it is essential during mouse embryogenesis where it is required for neural tube closure and other developmental events.[95]

Intriguingly, *lin-41* is not only a particularly well-studied target of *let-7*, but the LIN-41 protein might itself contribute to small RNA biogenesis and/or function as indicated by its co-immunoprecipitation with DCR-1 (Ref. 98), the pre-miRNA and dsRNA processing nuclease. No molecular function has yet been ascribed to LIN-41 in the process, and its loss does not appear to affect miRNA biogenesis.[98] However, the fly LIN-41 homologues Brat and Mei-P26 co-purify with Argonaute and, in the case of Mei-P26, do in fact appear to affect miRNA steady-state levels,[99] providing additional circumstantial support for LIN-41 in the process.

Other targets of *let-7* have been identified through various approaches, with a striking prevalence of transcription factors among the validated targets,[46,48,53,88–90] including *hbl-1* (Refs. 88 and 89), the *C. elegans* orthologue of the fly *hunchback* transcription factor, *daf-12* (Ref. 48), a nuclear hormone receptor and *pha-4* (Ref. 48), a forkhead transcription factor. Although the downstream effectors of these *let-7* target genes remain mostly unknown, it is likely that *let-7* indirectly coordinates the activity of a large number of genes through the regulation of these and other direct targets, probably even across multiple tissues. For instance, expression patterns and reporter gene analyses indicate that *let-7* regulates its targets *lin-41*, *daf-12*, *let-60/ras* and *K10C3.4* (a gene of unknown function) in the seam cells,[46,48,53] *hbl-1* in the ventral nerve cord,[88,89] *pha-4* in the intestine[48] and *K10C3.4* in the vulva.[46] It is a strong possibility that *let-7* may also regulate targets in yet other tissues,[90] as we do not know the full spatial expression pattern of *let-7* and most of its targets. Notably, however, *let-7* heterochronic phenotypes have so far only been investigated and confirmed in the seam cells (Figure 3.1a), suggesting that additional phenotypes and the targets that mediate them remain to be discovered. In this context, it also remains to be elucidated how genes regulated in the intestine or the ventral nerve cord can contribute to the vulval bursting phenotype of *let-7* mutant animals.

The retarded heterochronic phenotypes in the seam cells of *let-7* mutant animals, i.e. the failure to exit the cell cycle and differentiate,[54] are also hallmarks of cancer. Mammalian *let-7* has indeed now been validated as a potent tumour suppressor gene in lung, breast and possibly other epithelial cancers.[14] Elucidation of this function of mammalian *let-7* was a direct consequence of work in *C. elegans*, where a combined computational–experimental strategy identified *let-60*, the *C. elegans* orthologue of the human *RAS* proto-oncogenes as a target of *let-7* and its sister, *mir-84* (Refs. 48 and 100). *RAS* plays a prominent role in human lung cancers, and *in vitro* experiments subsequently showed that *let-7* could directly regulate the expression of human N-RAS and K-RAS proteins.[100] The observation that *let-7* levels were reduced in lung tumour samples relative to normal tissues and that conversely *RAS* was overexpressed in the tumours attested to the *in vivo* relevance of these findings,[100] which has now also been confirmed in mouse models.[101,102] As discussed in more detail elsewhere,[14] human *let-7* miRNAs (a family of 10 sequence-related miRNAs encoded by 13 genomic loci) regulate additional target genes, including *HMGA2*, *Myc* and *CDC25A*, further contributing to their tumour suppressive activity.

In *C. elegans*, consistent with its requirement for proper execution of the larval-to-adult transition, *let-7* RNA is abundant in L4 stage larvae and adults.[54,60,72] Weaker expression is already observed at the L3 stage[54,72] or, according to some reports, during the L2 stage,[60,61] but it is not known whether this reflects weak expression in all tissues where *let-7* is also expressed at the L4 stage, or strong expression in only a subset of the tissues. When the putative *let-7* promoter was used to drive expression of a GFP reporter, *let-7* transcription in the seam cells, the intestine and vulval precursor cells were found to be temporally regulated[3,72]; these are tissues where temporal regulation of the *let-7* targets has also been observed. The temporal regulation of *let-7* transcription is mediated by a promoter element of ~116 bp sequence termed 'temporal regulatory element' (TRE), which is both required and sufficient for this temporal expression profile in seam cells.[3] It contains a short inverted repeat element, possibly a binding site for an unidentified transcription factor.[3] Nonetheless, in many other tissues *let-7* promoter activity can already be seen in early larval and even embryonic stages,[15] indicating that post-transcriptional regulation further controls accumulation in at least some tissues and at some developmental stages, similar to what has been observed for mammalian *let-7* miRNAs.[14]

3.7.2 *Let-7 is part of a large microRNA family*

The *let-7* miRNA belongs to a group of seven related miRNAs in *C. elegans*.[79,103] These miRNAs are known as the '*let-7* family' because they share the so-called 'seed' sequence, i.e. nucleotides 2–8 of their 5′ ends,[79] which is important for target recognition (Section 3.4). Among the six *let-7* sisters, only *mir-48*, *mir-84* and *mir-241* have been studied whereas nothing is known about the expression patterns or developmental roles of the remaining three family members, *mir-793–mir-795*. Northern blot analysis revealed temporally regulated expressions of *mir-48*, *mir-84*, *mir-241*, with the highest levels seen in the L4 stage, as for *let-7*. However, expression generally precedes that of *let-7*, with *mir-84* first detectable at the early L1 stage, *mir-48* at the late L1 stage and *mir-241* at the L2 stage.[60,61,72] (Slightly different patterns were previously reported in Refs. 79 and 82.) Similarly to these related, yet divergent temporal expression profiles of the *let-7* family members, their spatial expression pattern also appears to overlap only partially. For instance, although all family members are expressed in the vulva, promoter activity is observed in distinct vulval precursor cells, i.e. P5.p–P7.p for *let-7*; P3.p–P8.p except for P6.p for *mir-84*; and P5.p and P7.p only for *mir-48* (Refs. 3 and 72).

Reflecting their prominent expression in the vulva and seam cells, *let-7* family members affect vulval and seam cell differentiation. Overexpression of *mir-84* from a transgenic array using its own promoter leads to precocious differentiation of vulval precursor and seam cells.[100] Similarly, a mutation in the promoter of *mir-48* (also known as *lin-58*) that causes precocious *mir-48* expression leads to weak precocious seam cell phenotypes.[61] Like *let-7*, the *let-7* sisters thus function as developmental timers. However, consistent with their earlier expression relative to *let-7*, *mir-48*, *mir-84* and *mir-241* regulate temporal patterning in the seam cells at an earlier developmental stage, i.e. the L2 stage (Figure 3.1a).[60]

The phenotypes seen upon overexpression of the *let-7* sisters[61,100] are not fully mirrored by the loss of function phenotypes. For instance, whereas overexpression of *mir-84* can suppress the multivulva (Muv) phenotype that results from *let-60/ras* hyperactivity,[100] deletion of *mir-84* does not cause a Muv phenotype.[60] Repression of *let-60/ras* by *mir-84* might thus be a neomorphic phenotype owing to non-physiological *mir-84* overexpression. Alternatively, additional activities, miRNA or protein, might provide redundant *let-60*/ras regulation that prevents a Muv phenotype even in the absence of *let-7* miRNAs. Indeed, although individual deletion of *mir-48*, *mir-84* and *mir-241* does not cause a strong developmental defect, their combined deletion results in a severe retarded heterochronic phenotype of the seam cells, although not the vulva.[60] Furthermore, overexpression of *mir-48* or mir-84 suppresses *let-7* phenotypes.[61,104] Finally, the *let-7* targets *hbl-1*, *daf-12* and *lin-41* are upregulated in *mir-48 mir-241*; *mir-84* triple mutant animals,[23,60] and *let-60/ras* has been identified as a target of both *let-7* and *mir-84* (Ref. 100).

Thus, although complementarity to only the miRNA seed sequence appears insufficient for regulation of mRNAs in *C. elegans* (Section 3.4; Refs. 40, 42 and 43), the *let-7* family members clearly have related functions not only at the developmental, but also at the molecular level. Nonetheless, their differences in temporal and inferred spatial expression patterns[3,60,61,72] suggest that *let-7* and its sister miRNAs regulate shared targets in different tissues, rather than being fully redundant. This notion is also supported by reporter gene analyses of *hbl-1*, which revealed regulation by *let-7* during late larval stages in the ventral nerve cord[88,89] and by the *let-7* sisters during early larval development in the hypodermal *hyp7* cell.[60]

An alternative, not mutually exclusive, model is that the genes regulated by *let-7* family members have different functions at different expression levels, as discussed above for the *lin-4* target *lin-14* (Section 3.6.1). In other words, the initial repression of *daf-12, lin-41, hbl-1* or *let-60* by the *let-7* sisters might reduce their levels sufficiently to abolish some of their functions, while still permitting others, which will only be terminated once further repression occurs through *let-7*. Intriguingly, the forkhead transcription factor PHA-4, a *let-7* target,[48] has been demonstrated to activate some promoters with high affinity binding sites at low expression levels, early in embryogenesis, whereas much higher expression permits subsequent binding and activation of promoters with low affinity binding sites later in embryogenesis.[105] A similarly graded response to PHA-4 during post-embryonic development appears likely and might in fact be a general feature of transcription factors, including those targeted by *let-7*.

3.8 Regulation of Neuronal Differentiation by *mir-273* and *lsy-6*

A second pathway that has contributed significantly to our understanding of miRNA function in development involves a pair of chemosensory asymmetric neurons: ASEL and ASER. The ASE neurons are symmetrically organised and look superficially identical, but they respond to distinct chemosensory cues, a functional asymmetry generated by a differential expression of a number of genes in the left ASEL and the right ASER neuron. Extensive genetic screens have identified a number of *lsy* mutations (loss of symmetry; 'lousy'), that cause failure of differentiation, and these include the miRNA *lsy-6* (Ref. 106). Reporter gene analysis indicated that the activity of the *lsy-6* promoter was limited to a small number of cells, mostly head neurons, including ASEL but not ASER. *Cog-1*, a homeobox transcription factor that functions in the ASE differentiation pathway, has a reciprocal expression to *lsy-6* in these neurons, i.e. it is active in ASER but repressed in ASEL, and this expression depends on its regulation by *lsy-6* (Figure 3.2). A second miRNA, *mir-273* shows the opposite expression pattern of *lsy-6*, that is, it is found in ASER but not

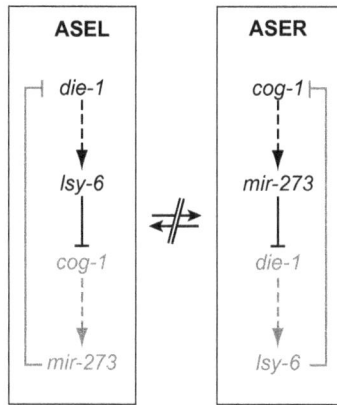

Figure 3.2   The miRNAs *lsy-6* and *mir-273* are selectively expressed in the asymmetric ASE left (ASEL) and ASE right (ASER) neurons, respectively. Whereas the *die-1* transcription factor promotes *lsy-6* expression in ASEL, *cog-1* promotes *mir-273* expression in ASER (possibly indirectly; *dashed* lines). By repressing *cog-1* in ASEL, *lsy-6* prevents expression of *mir-273*. Conversely, repression of *die-1* by *mir-273* prevents *lsy-6* expression in ASER. As a consequence, 'left' and 'right' identities are two mutually exclusive, stable fates. (*Black* indicates expression, *grey* lack of expression.)

ASEL.[107] In ASER, *mir-273* represses the activity of the zinc-finger transcription factor *die-1*, so that DIE-1 protein is only found in the ASEL, where it is required for the adoption of ASEL fate. A *mir-273* deletion mutant has recently become available, but has not yet been analysed for its function in ASE asymmetry.[80] The conclusion that asymmetric *mir-273* expression patterns are functionally important for ASE differentiation has thus largely relied on the demonstration that ectopic expression of *mir-273* in ASEL along with ASER prevents neuronal asymmetry.[107]

Strikingly, *mir-273* and *lsy-6* not only display reciprocal spatial expression patterns in ASEL/R, they also repress targets that are required for each other's expression. Thus, *mir-273* targets *die-1*, a positive regulator of *lsy-6* expression,[107] so that *lsy-6* is repressed in the presence of *mir-273*. Conversely, *lsy-6* represses *cog-1*, a positive regulator of *mir-273* expression,[106,107] so that *mir-273* expression requires the absence of *lsy-6*. The result is a double-negative feedback loop that achieves stable, mutually-exclusive expression patterns of these two miRNAs, ensuring stable cell fate switches (Figure 3.2).[108] However, it is not known if *die-1* and *cog-1* regulate transcription of their respective target miRNA directly, and it seems likely that this is indeed not the case.[108] It also remains unclear at what point this chain of events starts.[109]

3.9 *Mir-1*, a Highly Conserved, Muscle-Specific miRNA

Although several *C. elegans* miRNAs have readily recognisable counterparts in distant species including humans, perfect conservation of mature miRNA sequences is rare. Next to *let-7*, *mir-1* is the only other example of a miRNA that is perfectly conserved in sequence between *C. elegans* and humans, although it is one nucleotide shorter in *C. elegans*.[110] In mice and humans, miR-1 and its variant miR-206 are encoded by three different loci, *miR-1-1*, *miR-1-2* and *miR-206* (Ref. 110). Their muscle-specific expression is required for proper establishment and maintenance of muscle and cardiac tissue (reviewed in Ref. 111). Accordingly, mutations in *miR-1* result in early lethality due to defects in muscle proliferation in flies and mice.[112,113]

Given its sequence conservation and the essential function of miR-1 in various organisms, it was perplexing that *C. elegans mir-1* mutant animals are viable and exhibit apparently normal muscle development.[80,114] This does not seem to be due to redundancy, as no additional *mir-1* family members have been identified in *C. elegans*. Nonetheless, sequence conservation of *mir-1* does not appear to be a product of chance, as *mir-1* expression in *C. elegans* is restricted to muscles,[114] as in other organisms.[111] It has been speculated that the lack of an overt developmental phenotype in *C. elegans mir-1* mutants might reflect known differences in muscle development compared to fly and mouse, including differences in precursor cell proliferation and differentiation.[114]

Instead of functioning in muscle development, *C. elegans mir-1* was found to regulate both pre- and post-synaptic function at neuromuscular junctions.[114] Thus, *mir-1* mutant animals are resistant to treatment with levamisole, a chemical that stimulates the LevR subgroup of nicotinic acetylcholine receptors to cause muscle contraction and paralysis. In a beautiful set of experiments, Simon *et al.*[114] showed that this effect is, at least in part, mediated by *mir-1* influencing LevR subunit composition. *Mir-1* directly regulates the abundance of the UNC-29 and UNC-63 LevR subunits, and co-overexpression of these subunits in *mir-1*-expressing animals recapitulates the levamisole resistance phenotype seen with loss of *mir-1*.

In addition to regulating LevR, and thus post-synaptic activity, *mir-1* also appears to regulate the MEF-2 transcription factor.[114] MEF-2 activity in the muscle in turn controls a retrograde signal of unknown identity that regulates pre-synaptic acetylcholine release from the neuron. *Mir-1* thus has the potential to regulate and coordinate both pre-synaptic release of, and post-synaptic response to, this neurotransmitter, at least in response to external stimuli such as levamisole. Whether this reflects the physiological function of *mir-1* and how altered neurophysiology affects animal behaviour remains to be determined. It will also be interesting to find out whether this 'post-developmental' function of *mir-1* is conserved in other animals, where miR-1 has important developmental functions.

3.10 *MiR-61* as a Regulator of Vulval Differentiation

In addition to the miRNAs discussed above, *mir-61* is the only other miRNA that has received detailed scrutiny for its function in *C. elegans* development. Based on its expression pattern, its transcriptional regulation, the identification of a potential target, and ectopic expression phenotypes, a function in cell fate determination in the developing vulva has been inferred.[115]

Promoter analysis showed that *mir-61* is transcribed in the P5.p and P7.p vulval precursor cells, which are destined to adopt the so-called secondary vulval fate.[115] By contrast, no promoter activity was observed in the P6.p cell, destined to adopt the primary vulval fate. Ectopic *mir-61* expression in P6.p induced a secondary cell fate marker in this cell, supporting a function of orthotopically-expressed *mir-61* in this process in P5.p and P.7p.

One gene that might mediate this function of *mir-61* is *vav-1*, the *C. elegans* orthologue of the human proto-oncogene and putative cell signalling factor VAV, as the *vav-1* 3' UTR mediates repression of a reporter gene in P5.p and P7.p but not P6.p, an effect that depends on the predicted *mir-61* binding sites.[115] This finding is particularly intriguing because by genetic criteria *vav-1* functions as a repressor of *lin-12/notch*, whereas transcription of *mir-61* depends on *lin-12* (Ref. 115). Repression of *vav-1* by *mir-61* would thus complete a positive feedback loop resulting in sustained *lin-12* expression. Of note, however, as in the case of *mir-84*, a subsequent study of a deletion mutant of *mir-61* has failed to confirm its significance in vulval development inferred from the ectopic expression experiments.[80] Whether this is due to technical reasons (Section 3.11), redundancy with other regulatory pathways, or a truly neomorphic function resulting from ectopic or overexpression remains to be established and awaits more detailed dissection of the miRNA deletion strain.

3.11 Few miRNAs Appear to be Individually Important for *C. elegans* Development

The miRNAs discussed in detail above provide examples of miRNAs with defined, important roles in cellular function or development, but they make up only a tiny fraction of the currently 154 known *C. elegans* miRNAs[110] (miRBase Release 12.0; www.mirbase.org). An important question thus is whether the known functions of miRNAs are characteristic of miRNAs as a class, that is, will individual miRNAs typically have key regulatory roles in various developmental pathways?

A recent study[80] that reported deletion mutations in 87 miRNA genes in *C. elegans*, the first large-scale collection for any organism, provided strong arguments against this idea. Their results suggested that most miRNAs are, at least individually, dispensable. A first surprise was the finding that almost all miRNA deletion mutant animals were viable.[80] The exceptions were *mir-35~41*, in which case an entire cluster of sequence-related miRNAs was deleted and found to cause temperature-sensitive embryonic and larval lethality, and *mir-50* and *mir-353* deletions, both of which were inviable. However, in the case of *mir-50* and *mir-353*, impaired viability is likely due to the partial deletion of the respective 'host' genes (i.e. genes in whose introns these two miRNAs are located), as RNAi against either host genes, *rpl-24.1* for *mir-353* and *Y71G12B.11* for *mir-50*, is known to cause embryonic lethality.

Even more surprisingly, when Miska *et al.*[80] subjected their battery of miRNA deletion mutant animals to broad phenotypic assays, including an initial analysis of *C. elegans* morphology, growth, development and behaviour, only one additional mutation, deleting the *mir-240 mir-786* miRNA cluster, was found to result in a detectable phenotype, namely an abnormal defecation cycle length.

Although these results suggest that most miRNAs are not important for *C. elegans* development, several caveats need to be considered. First, almost none of the mutant strains investigated in the study had been back-crossed to wild-type animals prior to analysis to eliminate potentially confounding background mutations. This is important because genera-tion of *C. elegans* deletion mutants does not occur through targeted gene deletion, but through introduction of random deletions in the genome using mutagens. A collection of such mutants is screened for animals that harbour deletions in the gene of interest. Deletions might thus not only extend into neighbouring genes as described for *mir-50* and *mir-353* above but will also be accompanied by additional mutations, in unknown loca-tions, that could potentially modify miRNA deletion phenotypes, i.e. suppress them. Most notoriously, inversion or translocation of the targeted gene might provide continued gene expression despite the appearance of a gene deletion. However, although a major caveat when it comes to assessing the developmental roles (or lack thereof) of individual miRNAs, it appears unlikely that this fact could account for the general lack of detectable functions of most individual miRNAs.

A more relevant explanation appears to be that miRNAs frequently occur in 'families' of sequence-related miRNAs[79] that might have redun-dant functions, as exemplified by the *let-7* sister miRNAs *mir-48*, *mir-84* and *mir-241* (Section 3.7). Indeed, both mutants with newly-discovered phenotypes in these genomic studies[80] deleted more than one miRNA, although only *mir-35~41* constitute a 'family', whereas *mir-240* and *mir-786* are unrelated in sequence. Intriguingly, an eighth miRNA, *mir-42*, also belongs to the *mir-35~41* family and is located near the cluster although presumably transcribed independently of it. It will be interesting to see whether deletion of the entire family, i.e. the cluster and *mir-42*, will turn the temperature-sensitive lethal phenotype seen upon deletion of the cluster[80] into a constitutively lethal phenotype.

It also appears likely that additional miRNAs will have specific func-tions, but that the phenotypes resulting from their deletion will be either too subtle or too specific to be detectable in large-scale analyses such as that reported by Miska *et al.*,[80] which, by necessity, need to focus on rapid, relatively crude phenotypic analyses. As discussed above (Sections 3.8 and 3.9), *lsy-6* and *mir-1* are two miRNAs that, when deleted, result in very specific phenotypes. Altered neuronal differentiation in the *lsy-6* mutant is detectable using molecular markers or specific chemotaxis assays,[106] neither of which were employed in the genomics analysis. Similarly, whereas no overt phenotype was seen for *mir-1* deletion animals in this original analysis, Simon *et al.*[114] subsequently demonstrated its role in modulating synapse activity, which, however, only became visible in the presence of the acetylcholine receptor agonist levamisole. Thus, it appears likely that additional phenotypes will be detected in various miRNA deletion mutants once the relevant functions are tested with appropriate assays.

Nonetheless, further support for the notion that most *C. elegans* miRNAs cause only very subtle and/or specific phenotypes, at least under laboratory conditions, appears to be provided by the observation that animals lacking core components of the miRNA biogenesis machinery such as Dicer or Drosha still complete both embryogenesis and larval development.[6,10,11,116] These mutant animals display overt defects only at the larval-to-adult transition, namely the vulval bursting and seam cell defects characteristic of *let-7* mutations.[6,10,11] As a matter of fact, depletion of the *let-7* target *lin-41* can suppress these phenotypes partially, support- ing loss of *let-7* as a major cause of these phenotypes, when >100 additional miRNAs are affected. However, it has not yet been demon- strated that provision of mature *let-7* RNA can suppress *C. elegans* biogenesis machinery mutations. This caveat is worth keeping in mind as LIN-41 might itself function in small RNA pathways (Section 3.7).

Surviving Dicer- or Drosha-mutant larvae, i.e. those that do not suc- cumb to vulval bursting, grow into completely sterile adults.[6,10,11,98] There is thus a — direct or indirect — requirement of miRNAs for fertility, although it remains to be elucidated which miRNAs contribute when and how to this process. By contrast, the fact that both Dicer and Drosha dele- tion mutants complete embryonic and larval development would argue that miRNAs in general are largely dispensable for these processes. However, these homozygous mutant animals are derived from heterozy- gous mothers that might provide maternal RNA and/or protein. Thus, sufficient Dicer/Drosha protein activity might be available in the mutant animals for normal production of miRNAs with embryonic or early larval functions, but not of miRNAs such as *let-7* that only start to be expressed during late larval stages. Indeed, combined depletion of *alg-1* and *alg-2*, the two *C. elegans* miRISC Argonautes, results in synthetic embryonic and larval lethality,[11,117] directly confirming important function of miRNAs in early *C. elegans* development. The miRNAs required for early viability are not known, although one might expect the *mir-35~42* family[80] to be involved.

In summary, whereas most *C. elegans* miRNAs might not individually play the prominent developmental roles one would have come to expect from the examples of *lin-4* and *let-7*, lack of overt phenotypes appears to be largely a consequence of redundancy, and detection of mutant phenotypes might require more specific assays. We will consider in Section 3.12 how novel functions might be assigned to uncharacterised miRNAs.

3.12 Elucidating the Function of Uncharacterised miRNAs

The highly specific phenotypes seen upon deletion of some miRNAs suggests that knowledge of spatial and temporal miRNA expression patterns will be the key to assigning functions to uncharacterised miRNAs. Ideally, these patterns would be recorded for the mature miRNA and even highlight its activity, as miRNA biogenesis, and possibly activity, can be regulated.[13] Unfortunately, however, successful miRNA *in situ* hybridization in *C. elegans* has not yet been reported, so that these patterns remain unknown. Instead, northern blot analysis (or, more recently, deep-sequencing) of RNA recovered from whole worms of different developmental stages permits a crude analysis of temporal expression patterns, although changes in the signal could equally likely result from altered spatial domains of expression and altered expression levels in a fixed set of cells or tissues. By contrast, analysis of miRNA promoter activity by fusion of the putative miRNA promoter to a GFP reporter[3] can resolve expression patterns both temporally and spatially, and can distinguish between altered expression domains or levels. However, it will fail to take into account post-transcriptional regulation of miRNA accumulation and activity. Given these distinct shortcomings, these two approaches provide complementary tools that together can provide a reasonable approximation of miRNA expression patterns.

Several studies have gathered a large amount of data on both RNA levels and promoter activities of *C. elegans* miRNAs.[3,5,15,61,72,79,81,82,118] Northern blotting and small RNA cloning and sequencing have thus revealed that the majority of miRNAs are constitutively expressed during larval development,[79,81,82] although they might still be differentially expressed during embryogenesis or in specific tissues. Still, tens of miRNAs display expression patterns that change during larval development,[79,81,82,118] suggesting a role for these in mediating developmental functions. For most of these miRNAs, expression remains constant once it has been initiated, but the *mir-35~41* cluster, *mir-90*, *mir-247*, *mir-248* and *mir-234* are examples of miRNAs that are expressed in some stages and downregulated in subsequent stages.

How do these patterns compare to miRNA promoter activity, which has recently been recorded for a large fraction of *C. elegans* miRNA genes?[15] Before we answer this question, a few limitations of such studies need to be kept in mind. First, transgene arrays such as those used in these studies are typically silenced in the *C. elegans* germline so that failure to observe promoter activity in the germline might have technical reasons. Conversely, as in the case of protein, maternal contribution of RNA is possible, that is, miRNAs transcribed and matured in the mother might persist and be functional in the embryo, although no zygotic transcription occurs. Moreover, GFP used as a reporter in this study, is likely to have a different half-life from miRNAs and their precursors, affecting apparent promoter activity. On a temporal level, this issue can be, and in at least some instances[15] has been, ruled out to significantly affect the recorded patterns by quantifying the pri-miRNA levels using reverse-transcription quantitative PCR. Finally, no good tools exist for miRNA promoter prediction so that these analyses frequently rely on 'rules of thumb', using for instance the sequence upstream of a miRNA and up to the nearest neighbouring gene, or a defined sequence length as designated promoter. Only few 'rescue' experiments been reported, which can provide an approximation of the functionally relevant promoter size. Important regulatory elements might thus be missed. Given these potential limitations, it is surprising to learn that, generally, the promoter activities appeared to recapitulate quite faithfully the observed accumulation patterns of the miRNAs. Regulated transcription thus appears to provide a major control element for miRNA accumulation in *C. elegans*, although this might not be true for all tissues examined.

Recently, Martinez *et al.*[15] examined 70 miRNA promoters controlling expression of 89 miRNAs, among which they found 90% to confer GFP expression. For 39 out of 60 of these promoters (65%), GFP expression patterns were consistent with the known miRNA accumulation patterns, whereas 17 promoters (28%) exhibited only partially consistent, and 4 (7%) inconsistent expression patterns. These latter two classes might thus comprise incomplete promoters, lacking positive or negative regulatory elements, or, more interestingly, contain instances of post-transcriptionally regulated miRNAs. Consistent with the latter explanation, some miRNAs were identified that are assumed to be co-transcribed as clusters, i.e. from the same promoter, but nonetheless exhibit divergent expression patterns of the mature miRNA.[15] Examples of this include the *mir-61~250* cluster and the *mir-42~44* cluster. Other miRNAs with putative post-transcriptional regulation are the *mir-35~41* cluster, whose promoter is active, and pri-miRNA detectable, in early larval stages when little or no mature miRNA accumulates, and *let-7*, for which expression by northern blot is first detectable in L3 larvae,[54,72] whereas its promoter is already active in embryos as judged by both detection of a GFP signal and endogenous pri-*let-7* (Ref. 15). The extensive post-transcriptional regulation of *let-7* observed in mammals[13] might thus also occur on *C. elegans let-7*, although it remains to be established whether similar mechanisms are involved.

The availability of this large set of miRNA promoter activity patterns[15] has also enabled some more general analyses. In particular, it emerged that miRNA promoters are active in all major tissues and cell types (except for the germinal gonad, probably due to the technical reasons discussed above). Most miRNA promoters confer expression only in a few tissues or cell types, frequently three or fewer tissues, whereas only *lin-4*, *let-7* and *mir-53* display near-ubiquitous somatic expression although, at least for *lin-4* and *let-7*, expression still appears to be temporally regulated and is not seen in all cells.[3,72] Notably, although miRNAs from the same family do not necessarily exhibit identical spatial patterns of promoter activity, the overlap of these tissue expression patterns are more similar than those seen for miRNAs from distinct families.[15] This finding might lend further support to the notion that redundancy among family members is why most miRNAs appear individually dispensable. However, as shown for the *let-7* family members, all of which appear to be expressed in the vulva but, at least in part, in non-overlapping cells within the vulva,[72] 'tissue' might still be a fairly crude category when assigning spatial expression patterns.

An intriguing finding from the study by Martinez *et al.*[15] is that 'intronic miRNAs' that reside in the introns of 'host genes', a rare category in *C. elegans* where most miRNAs are encoded intergenically, may in fact be independent transcripts, rather than 'guests' of their host genes. Transgenes using the immediate upstream sequence of four intragenic miRNAs (*mir-58*, embedded in an intron of *Y67D8A.1*; *mir-2* and *mir-71*, both within the same intron of *ppfr-1* but separated by approximately 7 kb of sequence; and *mir-82*, in an intron of *T07D1.2*) revealed promoter activity for *mir-2*, *mir-71* and *mir-82* but not *mir-58* upstream sequences. Additionally, *lin-4* was initially presumed to be intergenic, but more recent data has found it to be located in an intron of *F59G1.4*. Nonetheless, the sequence in this intron not only drives expression of a GFP reporter[4,72] but also suffices to rescue *lin-4* null mutations when used to drive *lin-4* from a transgenic array,[65] unequivocally demonstrating that *lin-4* expression occurs independently of the host gene promoter. In contrast to *C. elegans*, most human miRNAs are located within introns.[119] Given the huge size of many human introns, it will be particularly interesting to see if the genomic sequences upstream of these miRNAs can similarly function as promoters.

Returning to the issue of how we can assign functions to uncharacterised miRNAs, the large set of miRNA expression patterns and the collection of miRNA deletion mutants[80] will be valuable resources to do just that. As promoter activity appears to have significant predictive value for miRNA accumulation patterns, and with most miRNAs appearing to have spatially restricted expression patterns, these expression patterns will guide us in which phenotypes to expect in which tissues from miRNA deletions, so that more specific and sensitive assays can be used on the mutant strains. Expression patterns have indeed been key to assigning functions to *mir-84, mir-61* and *mir-273* (see relevant sections above).

Moreover, as in other systems, *C. elegans* miRNAs have been found to reduce the levels of their target mRNAs,[22,23] as well as block their translation so that microarray analysis can be brought to bear on the mutant strains. Although such experiments will always pick up a large number of secondary targets, deregulated for instance because miRNAs directly repress transcription factors, this might indeed be beneficial for such analysis as it can help to infer entire deregulated pathways, providing a strong clue to biological function.

3.13 MiRNA Regulatory Networks

The fact that miRNA expression patterns are frequently dynamic and/or restricted to only certain tissues (e.g. Refs. 15 and 79) has also sparked an interest in how these expression patterns are achieved, and in particular in the identity of the transcriptional inputs that direct them. Using a yeast-1-hybrid screen, Walhout *et al.*[5] identified transcription factors binding to miRNA promoters and in this process identified 347 high-confidence interactions between 63 miRNA promoters and 116 proteins. Some miRNA promoters, including those of the *let-7* and *lin-4* families, were bound by several transcription factors, which might reflect the important developmental functions of these miRNAs and/or their expression in many tissues. Notably, *let-7* itself also appears to regulate many transcription factors as direct targets, suggesting the possibility of feedback loops. Indeed, when various miRNA target prediction programs were used to infer targeted transcription factors, 73 transcription factors were predicted to be regulated by 67 miRNAs resulting in 252 predicted interactions. These included 23 predicted miRNA–transcription factor feedback loops, in which a miRNA is predicted to regulate the same transcription factor that binds its promoter.[5] As computational miRNA target prediction is notoriously noisy, these predictions need to be confirmed experimentally; however, this finding is intriguing because it provides a simple regulatory module to achieve miRNA and/or transcription factor homeostasis. Indeed, examples of indirect feedback loops have been reported for *let-7*, *lsy-6/mir-273* and *mir-61* (Sections 3.7, 3.8 and 3.10), and might therefore constitute a common feature of miRNA pathways. The finding that many of the miRNAs and transcription factors that participate in these feedback loops are also highly connected to other transcription factors and miRNAs, respectively, then suggests the possibility of a highly adaptable regulation of these loops, which might provide an additional explanation for why disruption of individual miRNAs is frequently tolerable.[80]

3.14 Conclusions and Outlook

The depletion of miRNAs through interference with their biogenesis or functional machinery has confirmed the importance of these small RNAs for *C. elegans* development, and genetic and other approaches have assigned functions to miRNAs involving regulation of cell differentiation, proliferation or neuronal activity. Notably, at least some of these functions appear to be conserved in their mammalian counterparts. However, the challenge for the coming years will be to identify the functions of the large majority of miRNAs where these have remained unknown. The availability of unique resources including a large library of miRNA deletion mutant animals[80] and knowledge of miRNA promoter activities[15] will help to achieve this goal. They might also help to address additional, more systemic issues such as the composition and functions of regulatory networks in which miRNAs are embedded. In this context, the recent realisation that miRNAs themselves are highly regulated, at the transcriptional as well as at the post-transcriptional level, highlights a new challenge, that is, identification of the mechanisms that regulate the miRNAs. It is not difficult to predict that *C. elegans* genetics, cell biology and genomics will make important contributions to these fields, as they have been doing for our general understanding of miRNA biology for more than 15 years.

Acknowledgments

We thank Aurora Esquela-Kerscher for sharing unpublished data and comments on the manuscript. We also thank Benjamin Hurschler and Ingo Büssing for critical readings of the manuscript. Work in H.G.'s lab is supported by the Novartis Research Foundation, the Swiss National Science Foundation and an ERC Starting Independent Researcher Grant.

References

1. Kim V.N. (2005). MicroRNA biogenesis: Coordinated cropping and dicing. *Nat Rev Mol Cell Biol* **6**, 376–385.
2. Bracht J., Hunter S., Eachus R. *et al.* (2004). Trans-splicing and polyadenylation of let-7 microRNA primary transcripts. *RNA* **10**, 1586–1594.
3. Johnson S.M., Lin S.Y. and Slack F.J. (2003). The time of appearance of the *C. elegans* let-7 microRNA is transcriptionally controlled utilizing a temporal regulatory element in its promoter. *Dev Biol* **259**, 364–379.
4. Ow M.C., Martinez N.J., Olsen P.H. *et al.* (2008).The FLYWCH transcription factors FLH-1, FLH-2, and FLH-3 repress embryonic expression of microRNA genes in *C. elegans. Genes Dev* **22**, 2520–2534.
5. Martinez N.J., Ow M.C., Barrasa M.I. *et al.* (2008). A *C. elegans* genome-scale microRNA network contains composite feedback motifs with high flux capacity. *Genes Dev* **22**, 2535–2549.
6. Denli A.M., Tops B.B., Plasterk R.H. *et al.* (2004). Processing of primary microRNAs by the Microprocessor complex. *Nature* **432**, 231–235.
7. Okamura K., Hagen J.W., Duan H. *et al.* (2007). The mirtron pathway generates microRNA-class regulatory RNAs in Drosophila. *Cell* **130**, 89–100.
8. Berezikov E., Chung W.J., Willis J. *et al.* (2007). Mammalian mirtron genes. *Mol Cell* **28**, 328–336.
9. Ruby J.G., Jan C.H. and Bartel D.P. (2007). Intronic microRNA precursors that bypass Drosha processing. *Nature* **448**, 83–86.
10. Ketting R.F., Fischer S.E., Bernstein E. *et al.* (2001). Dicer functions in RNA interference and in synthesis of small RNA involved in developmental timing in *C. elegans. Genes Dev* **15**, 2654–2659.
11. Grishok A., Pasquinelli A.E., Conte D. *et al.* (2001). Genes and mechanisms related to RNA interference regulate expression of the small temporal RNAs that control *C. elegans* developmental timing. *Cell* **106**, 23–34.
12. Hutvagner G., Simard M.J., Mello C.C. *et al.* (2004). Sequence-specific inhibition of small RNA function. *PLoS Biol* **2**, E98.
13. Ding X.C., Weiler J. and Großhans H. (2009). Regulating the regulators: Mechanisms controlling the maturation of microRNAs. *Trends Biotechnol* **27**, 27–36.
14. Büssing I., Slack F.J. and Großhans H. (2008). Let-7 microRNAs in development, stem cells and cancer. *Trends Mol Med* **14**, 400–409.
15. Martinez N.J., Ow M.C. Reece-Hoyes J.S *et al.* (2008). Genome-scale spatiotemporal analysis of *Caenorhabditis elegans* microRNA promoter activity. *Genome Res* **18**, 2005–2015.

16. Bohnsack M.T., Czaplinski K. and Gorlich D. (2004). Exportin 5 is a RanGTP-dependent dsRNA-binding protein that mediates nuclear export of pre-miRNAs. *RNA* **10**, 185–191.

17. Murphy D., Dancis B. and Brown J.R. (2008). The evolution of core proteins involved in microRNA biogenesis. *BMC Evol Biol* **8**, 92.

18. Blumenthal T. (2005). *Trans-splicing and operons.* [Online]. Available at: http://www.wormbook.org/chapters/www_transsplicingoperons/transsplicing-operons.html.

19. Filipowicz W., Bhattacharyya S.N. and Sonenberg N. (2008). Mechanisms of post-transcriptional regulation by microRNAs: Are the answers in sight? *Nat Rev Genet* **9**, 102–114.

20. Olsen P.H. and Ambros V. (1999). The *lin-4* regulatory RNA controls developmental timing in *Caenorhabditis elegans* by blocking LIN-14 protein synthesis after the initiation of translation. *Dev Biol* **216**, 671–680.

21. Seggerson K., Tang L. and Moss E.G. (2002). Two genetic circuits repress the *Caenorhabditis elegans* heterochronic gene lin-28 after translation initiation. *Dev Biol* **243**, 215–225.

22. Bagga S., Bracht J., Hunter S. *et al.* (2005). Regulation by let-7 and *lin-4* miRNAs results in target mRNA degradation. *Cell* **122**, 553–563.

23. Ding X.C. and Großhans H. (2009). Repression of *C. elegans* microRNA targets at the initiation level of translation requires GW182 proteins. *EMBO J* **28**, 213–222.

24. Holtz J. and Pasquinelli A.E. (2009). Uncoupling of *lin-14* mRNA and protein repression by nutrient deprivation in *Caenorhabditis elegans. RNA* **15**, 400–405.

25. Ding L., Spencer A., Morita K. *et al.* (2005). The developmental timing regulator AIN-1 interacts with miRISCs and may target the argonaute protein ALG-1 to cytoplasmic P bodies in *C. elegans. Mol Cell* **19**, 437–447.

26. Zhang L., Ding L., Cheung T.H. *et al.* (2007). Systematic identification of *C. elegans* miRISC proteins, miRNAs, and mRNA targets by their interactions with GW182 proteins AIN-1 and AIN-2. *Mol Cell* **28**, 598–613.

27. Chendrimada T.P., Finn K.J., Ji X. *et al.* (2007). MicroRNA silencing through RISC recruitment of eIF6. *Nature* **447**, 823–828.

28. Eulalio A., Rehwinkel J., Stricker M. *et al.* (2007). Target-specific requirements for enhancers of decapping in miRNA-mediated gene silencing. *Genes Dev* **21**, 2558–2570.

29. Eulalio A., Huntzinger E. and Izaurralde E. (2008). GW182 interaction with Argonaute is essential for miRNA-mediated translational repression and mRNA decay. *Nat Struct Mol Biol* **15**, 346–353.

30. Gandin V., Miluzio A., Barbieri A.M. *et al.* (2008). Eukaryotic initiation factor 6 is rate-limiting in translation, growth and transformation. *Nature* **455**, 684–688.

31. Ding X.C., Slack F.J. and Großhans H. (2008). The let-7 microRNA interfaces extensively with the translation machinery to regulate cell differentiation. *Cell Cycle* **7**, 3083–3090.

32. Caudy A.A., Myers M., Hannon G.J. *et al.* (2002). M. Fragile X-related protein and VIG associate with the RNA interference machinery. *Genes Dev* **16**, 2491–2496.

33. Chan S.P., Ramaswamy G., Choi E.Y. *et al.* (2008). Identification of specific let-7 microRNA binding complexes in *Caenorhabditis elegans*. *RNA* **14**, 2104–2114.

34. Parry D.H., Xu J. and Ruvkun G.A. (2007). Whole-genome RNAi screen for *C. elegans* miRNA pathway genes. *Curr Biol* **17**, 2013–2022.

35. Cai Q., Sun Y., Huang X. *et al.* (2008). The *C. elegans* PcG-like gene sop-2 regulates the temporal and sexual specificities of cell fates. *Genetics* **178**, 1445–1456.

36. Pillai R.S., Artus C.G. and Filipowicz W. (2004). Tethering of human Ago proteins to mRNA mimics the miRNA-mediated repression of protein synthesis. *RNA* **10**, 1518–1525.

37. Behm-Ansmant I., Rehwinkel J., Doerks T. *et al.* (2006). mRNA degradation by miRNAs and GW182 requires both CCR4:NOT deadenylase and DCP1:DCP2 decapping complexes. *Genes Dev* **20**, 1885–1898.

38. Vella M.C., Reinert K. and Slack F.J. (2004). Architecture of a validated microRNA::target interaction. *Chem Biol* **11**, 1619–1623.

39. Rajewsky N. (2006). MicroRNA target predictions in animals. *Nat Genet* **38 Suppl**, S8–S13.

40. Didiano D. and Hobert O. (2006). Perfect seed pairing is not a generally reliable predictor for miRNA-target interactions. *Nat Struct Mol Biol* **13**, 849–851.

41. Ha I., Wightman B. and Ruvkun G.A. (1996). Bulged *lin-4*/lin-14 RNA duplex is sufficient for *Caenorhabditis elegans* lin-14 temporal gradient formation. *Genes Dev* **10**, 3041–3050.

42. Vella M.C., Choi E.Y., Lin S.Y. *et al.* (2004). The *C. elegans* microRNA let-7 binds to imperfect let-7 complementary sites from the lin-41 3′ UTR. *Genes Dev* **18**,132–137.

43. Didiano D. and Hobert O. (2008). Molecular architecture of a miRNA-regulated 3′ UTR. *RNA* **14**, 1297–1317.

44. Long D., Lee R., Williams P. *et al.* (2007). Potent effect of target structure on microRNA function. *Nat Struct Mol Biol* **14**, 287–294.

45. Kertesz M., Iovino N., Unnerstall U. *et al.* (2007). The role of site accessibility in microRNA target recognition. *Nat Genet* **39**, 1278–1284.
46. Andachi Y. (2008). A novel biochemical method to identify target genes of individual microRNAs: Identification of a new *Caenorhabditis elegans* let-7 target. *RNA* **14**, 2440–2451.
47. Hammell M., Long D., Zhang L. *et al.* (2008). mirWIP: MicroRNA target prediction based on microRNA-containing ribonucleoprotein-enriched transcripts. *Nat Methods* **5**, 813–819.
48. Großhans H., Johnson T., Reinert K.L. *et al.* (2005). The temporal patterning microRNA let-7 regulates several transcription factors at the larval to adult transition in *C. elegans*. *Dev Cell* **8**, 321–330.
49. Bartel D.P. and Chen C.Z. (2004). Micromanagers of gene expression: The potentially widespread influence of metazoan microRNAs. *Nat Rev Genet* **5**, 396–400.
50. Ruvkun G., Wightman B., Burglin T. *et al.* (1991). Dominant gain-of-function mutations that lead to misregulation of the *C. elegans* heterochronic gene lin-14, and the evolutionary implications of dominant mutations in pattern-formation genes. *Dev Suppl* **1**, 47–54.
51. Ambros V. (1989). A hierarchy of regulatory genes controls a larva-to-adult developmental switch in *C. elegans*. *Cell* **57**, 49–57.
52. Arasu P., Wightman B. and Ruvkun G. (1991). Temporal regulation of lin-14 by the antagonistic action of two other heterochronic genes, lin-4 and lin-28. *Genes Dev* **5**, 1825–1833.
53. Slack F.J., Basson M., Liu Z. *et al.* (2000). The lin-41 RBCC gene acts in the *C. elegans* heterochronic pathway between the let-7 regulatory RNA and the LIN-29 transcription factor. *Mol Cell* **5**, 659–669.
54. Reinhart B.J., Slack F.J., Basson M. *et al.* (2000). The 21-nucleotide let-7 RNA regulates developmental timing in *Caenorhabditis elegans*. *Nature* **403**, 901–906.
55. Moss E.G. (2007). Heterochronic genes and the nature of developmental time. *Curr Biol* **17**, R425–R434.
56. Banerjee D. and Slack F.J. (2002). Control of developmental timing by small temporal RNAs: A paradigm for RNA-mediated regulation of gene expression. *Bioessays* **24**, 119–129.
57. Sulston J.E. and Horvitz H.R. (1977). Post-embryonic cell lineages of the nematode, *Caenorhabditis elegans*. *Dev Biol* **56**, 110–156.
58. Sulston J.E., Schierenberg E., White J.G. *et al.* (1983). The embryonic cell lineage of the nematode *Caenorhabditis elegans*. *Dev Biol* **100**, 64–119.

59. Chalfie M., Horvitz H.R. and Sulston J.E. (1981). Mutations that lead to reiterations in the cell lineages of *C. elegans*. *Cell* **24**, 59–69.

60. Abbott A.L., Alvarez-Saavedra E., Miska E.A. *et al.* (2005). The let-7 MicroRNA family members mir-48, mir-84, and mir-241 function together to regulate developmental timing in *Caenorhabditis elegans*. *Dev Cell* **9**, 403–414.

61. Li M., Jones-Rhoades M.W., Lau N.C. *et al.* (2005). Regulatory mutations of *mir-48*, a *C. elegans* let-7 family MicroRNA, cause developmental timing defects. *Dev Cell* **9**, 415–422.

62. Ambros V. and Horvitz H.R. (1984). Heterochronic mutants of the nematode *Caenorhabditis elegans*. *Science* **226**, 409–416.

63. Ambros V. and Horvitz H.R. (1987). The *lin-14* locus of *Caenorhabditis elegans* controls the time of expression of specific postembryonic developmental events. *Genes Dev* **1**, 398–414.

64. Ruvkun G., Ambros V., Coulson A. *et al.* (1989). Molecular genetics of the *Caenorhabditis elegans* heterochronic gene *lin-14*. *Genetics* **121**, 501–516.

65. Lee R.C., Feinbaum R.L. and Ambros V. (1993). The *C. elegans* heterochronic gene *lin-4* encodes small RNAs with antisense complementarity to lin-14. *Cell* **75**, 843–854.

66. Wightman B., Ha I. and Ruvkun G. (1993). Posttranscriptional regulation of the heterochronic gene lin-14 by lin-4 mediates temporal pattern formation in *C. elegans*. *Cell* **75**, 855–862.

67. Feinbaum R. and Ambros V. (1999). The timing of *lin-4* RNA accumulation controls the timing of postembryonic developmental events in *Caenorhabditis elegans*. *Dev Biol* **210**, 87–95.

68. Ruvkun G. and Giusto J. (1989). The *Caenorhabditis elegans* heterochronic gene lin-14 encodes a nuclear protein that forms a temporal developmental switch. *Nature* **338**, 313–319.

69. Reinhart B. J. and Ruvkun G. (2001). Isoform-specific mutations in the *Caenorhabditis elegans* heterochronic gene lin-14 affect stage-specific patterning. *Genetics* **157**, 199–209.

70. Hong Y., Lee R. C. and Ambros V. (2000). Structure and function analysis of LIN-14, a temporal regulator of postembryonic developmental events in *Caenorhabditis elegans*. *Mol Cell Biol* **20**, 2285–2295.

71. Hristova M., Birse D., Hong Y. *et al.* (2005). The *Caenorhabditis elegans* heterochronic regulator LIN-14 is a novel transcription factor that controls the developmental timing of transcription from the insulin/insulin-like growth factor gene ins-33 by direct DNA binding. *Mol Cell Biol* **25**, 11059–11072.

72. Esquela-Kerscher A., Johnson S.M. Bai L. *et al.* (2005). Post-embryonic expression of *C. elegans* microRNAs belonging to the *lin-4* and let-7 families in the hypodermis and the reproductive system. *Dev Dyn* **234**, 868–877.

73. Baugh L.R. and Sternberg P.W. (2006). DAF-16/FOXO regulates transcription of cki-1/Cip/Kip and repression of lin-4 during *C. elegans* L1 arrest. *Curr Biol* **16**, 780–785.

74. Liu Z. and Ambros V. (1991). Alternative temporal control systems for hypo-dermal cell differentiation in *Caenorhabditis elegans*. *Nature* **350**, 162–165.

75. Moss E.G., Lee R.C. and Ambros V. (1997). The cold shock domain protein LIN-28 controls developmental timing in *C. elegans* and is regulated by the *lin-4* RNA. *Cell* **88**, 637–646.

76. Morita K. and Han M. (2006). Multiple mechanisms are involved in regulating the expression of the developmental timing regulator lin-28 in *Caenorhabditis elegans*. *EMBO J* **25**, 5794–5804.

77. Rybak A., Fuchs H., Smirnova L. *et al.* (2008). A feedback loop comprising lin-28 and let-7 controls pre-let-7 maturation during neural stem-cell com-mitment. *Nat Cell Biol* **10**, 987–993.

78. Boehm M. and Slack F.J. (2005). A developmental timing microRNA and its target regulate life span in *C. elegans*. *Science* **310**, 1954–1957.

79. Lim L.P., Lau N.C., Weinstein E.G. *et al.* (2003). The microRNAs of *Caenorhabditis elegans*. *Genes Dev* **17**, 991–1008.

80. Miska E.A., Alvarez-Saavedra E., Abbott A. L. *et al.* (2007). Most *Caenorhabditis elegans* microRNAs are individually not essential for development or viability. *PLoS Genet* **3**, e215.

81. Lee R. C. and Ambros V. (2001). An extensive class of small RNAs in *Caenorhabditis elegans*. *Science* **294**, 862–864.

82. Lau N.C., Lim L.P., Weinstein E.G. *et al.* (2001). An abundant class of tiny RNAs with probable regulatory roles in *Caenorhabditis elegans*. *Science* **294**, 858–862.

83. Lagos-Quintana M., Rauhut R., Lendeckel W. *et al.* (2001). Identification of novel genes coding for small expressed RNAs. *Science* **294**, 853–858.

84. Lagos-Quintana M., Rauhut R., Yalcin A. *et al.* (2002). Identification of tissue-specific microRNAs from mouse. *Curr Biol* **12**, 735–739.

85. Sempere L.F., Freemantle S., Pitha-Rowe I. *et al.* (2004). Expression profiling of mammalian microRNAs uncovers a subset of brain-expressed microRNAs with possible roles in murine and human neuronal differentiation. *Genome Biol* **5**, R13.

86. Yu J., Vodyanik M.A., Smuga-Otto K. *et al.* (2007). Induced pluripotent stem cell lines derived from human somatic cells. *Science* **318**, 1917–1920.

87. Pasquinelli A.E., Reinhart B.J., Slack F. *et al.* (2000). Conservation of the sequence and temporal expression of let-7 heterochronic regulatory RNA. *Nature* **408**, 86–89.

88. Abrahante J.E., Daul A.L., Li M. *et al.* (2003). The *Caenorhabditis elegans* hunchback-like gene lin-57/hbl-1 controls developmental time and is regulated by microRNAs. *Dev Cell* **4**, 625–637.

89. Lin S.Y., Johnson S.M., Abraham M. *et al.* (2003). The *C. elegans* hunchback homolog, hbl-1, controls temporal patterning and is a probable microRNA target. *Dev Cell* **4**, 639–650.

90. Lall S., Grün D., Krek A. *et al.* (2006). A genome-wide map of conserved microRNA targets in *C. elegans*. *Curr Biol* **16**, 460–471.

91. Lowe S.W., Cepero E. and Evan G. (2004). Intrinsic tumour suppression. *Nature* **432**, 307–315.

92. Lancman J.J., Caruccio N.C., Harfe B.D. *et al.* (2005). Analysis of the regulation of *lin-41* during chick and mouse limb development. *Dev Dyn* **234**, 948–960.

93. Schulman B.R., Esquela-Kerscher A. and Slack F.J. (2005). Reciprocal expression of lin-41 and the microRNAs let-7 and mir-125 during mouse embryogenesis. *Dev Dyn* **234**, 1046–1054.

94. Kanamoto T., Terada K., Yoshikawa H. *et al.* (2006). Cloning and regulation of the vertebrate homologue of lin-41 that functions as a heterochronic gene in *Caenorhabditis elegans*. *Dev Dyn* **235**, 1142–1149.

95. Maller Schulman B.R., Liang X., Stahlhut C. *et al.* (2008). The let-7 microRNA target gene, Mlin41/Trim71 is required for mouse embryonic survival and neural tube closure. *Cell Cycle* **7**, 3935–3942.

96. Kloosterman W.P., Wienholds E., Ketting R.F. *et al.* (2004). Substrate requirements for let-7 function in the developing zebrafish embryo. *Nucl Acids Res* **32**, 6284–6291.

97. Lin Y.C., Hsieh L.C., Kuo M.W. *et al.* (2007). Human TRIM71 and its nematode homologue are targets of let-7 microRNA and its zebrafish orthologue is essential for development. *Mol Biol Evol* **24**, 2525–2534.

98. Duchaine T.F., Wohlschlegel J.A., Kennedy S. *et al.* (2006). Functional proteomics reveals the biochemical niche of *C. elegans* DCR-1 in multiple small-RNA-mediated pathways. *Cell* **124**, 343–354.

99. Neumüller R.A., Betschinger J., Fischer A. *et al.* (2008). Mei-P26 regulates microRNAs and cell growth in the Drosophila ovarian stem cell lineage. *Nature* **454**, 241–245.

100. Johnson S.M., Grosshans H., Shingara J. *et al.* (2005). RAS is regulated by the let-7 microRNA family. *Cell* **120**, 635–647.

101. Esquela-Kerscher A., Trang P., Wiggins J.F. *et al.* (2008). The let-7 microRNA reduces tumor growth in mouse models of lung cancer. *Cell Cycle* 7, 759–764.

102. Kumar M.S., Erkeland S.J., Pester R.E. *et al.* (2008). Suppression of non-small cell lung tumor development by the let-7 microRNA family. *Proc Natl Acad Sci USA* 105, 3903–3908.

103. Ruby J.G., Jan C., Player C. *et al.* (2006). Large-scale sequencing reveals 21U-RNAs and additional microRNAs and endogenous siRNAs in *C. elegans. Cell* 127, 1193–1207.

104. Hayes G.D., Frand A.R. and Ruvkun G. (2006). The mir-84 and let-7 paralogous microRNA genes of *Caenorhabditis elegans* direct the cessation of molting via the conserved nuclear hormone receptors NHR-23 and NHR-25. *Development* 133, 4631–4641.

105. Gaudet J., Muttumu S., Horner M. *et al.* (2004). Whole-genome analysis of temporal gene expression during foregut development. *PLoS Biol* 2, e352.

106. Johnston R.J. and Hobert O. (2003). A microRNA controlling left/right neuronal asymmetry in *Caenorhabditis elegans. Nature* 426, 845–849.

107. Chang S., Johnston R.J., Frøkjaer-Jensen C. *et al.* (2004). MicroRNAs act sequentially and asymmetrically to control chemosensory laterality in the nematode. *Nature* 430, 785–789.

108. Johnston R.J., Chang S., Etchberger J.F. *et al.* (2005). MicroRNAs acting in a double-negative feedback loop to control a neuronal cell fate decision. *Proc Natl Acad Sci USA* 102, 12449–12454.

109. Johnston R.J.J. and Hobert O. (2005). A novel *C. elegans* zinc finger transcription factor, lsy-2, required for the cell type-specific expression of the lsy-6 microRNA. *Development* 132, 5451–5460.

110. Griffiths-Jones S., Saini H.K., Dongen S.V. *et al.* (2007). miRBase: Tools for microRNA genomics. *Nucl Acids Res* 36, D154–D158.

111. Niu Z., Li A., Zhang S.X. *et al.* (2007). Serum response factor micromanaging cardiogenesis. *Curr Opin Cell Biol* 19, 618–627.

112. Sokol N.S. and Ambros V. (2005). Mesodermally expressed Drosophila microRNA-1 is regulated by Twist and is required in muscles during larval growth. *Genes Dev* 19, 2343–2354.

113. Zhao Y., Samal E. and Srivastava D. (2005). Serum response factor regulates a muscle-specific microRNA that targets Hand2 during cardiogenesis. *Nature* 436, 214–220.

114. Simon D.J., Madison J.M., Conery A.L. *et al.* (2008). The microRNA miR-1 regulates a MEF-2-dependent retrograde signal at neuromuscular junctions. *Cell* 133, 903–915.

115. Yoo A. S. and Greenwald I. (2005). LIN-12/Notch activation leads to microRNA-mediated downregulation of Vav in *C. elegans*. *Science* **310**, 1330–1333.

116. Knight S.W. and Bass B.L. (2001). A role for the RNase III enzyme DCR-1 in RNA interference and germ line development in *Caenorhabditis elegans*. *Science* **293**, 2269–2271.

117. Tops B.B., Plasterk R.H. and Ketting R.F. (2006). The *Caenorhabditis elegans* Argonautes ALG-1 and ALG-2: Almost identical yet different. *Cold Spring Harb Symp Quant Biol* **71**, 189–194.

118. Gu S.G., Pak J., Barberan-Soler S. *et al.* (2007). Distinct ribonucleoprotein reservoirs for microRNA and siRNA populations in *C. elegans*. *RNA* **13**, 1–13.

119. Kim Y.K. and Kim V.N. (2007). Processing of intronic microRNAs. *EMBO J* **26**, 775–783.

4

MicroRNAs in Mammalian Development

*Andrea Ventura**

Despite the significant advances in understanding their biogenesis and mechanism of action, we still know relatively little with respect to the exact biological functions of the vast majority of mammalian microRNAs. Fortunately, this is rapidly changing, at least in part thanks to the use of sophisticated mouse models. In particular, ablation of microRNA biogenesis at specific developmental stages and in a tissue-restricted fashion is revealing that although miRNAs are largely dispensable for organogenesis, they are essential to ensure terminal differentiation and tissue homeostasis in a number of different settings.

In addition, the characterisation of constitutive and conditional loss-of–function and gain-of-function alleles of individual microRNAs and microRNA clusters is providing a wealth of data regarding their specific functions *in vivo*, the targets they modulate, and the degree of functional overlap existing between related microRNA genes.

*Cancer Biology and Genetics Programme, Memorial Sloan-Kettering Cancer Centre, 408 East 69th, ZRC-1201, New York, NY 10065, USA. E-mail: venturaa@mskcc.org.

4.1 Introduction

Since their discovery 16 years ago, miRNA have emerged as critical regulators of gene expression in metazoans and in plants. As we write, over
700 human miRNAs have been identified, a number that is destined to
increase. While a great deal is known about their biogenesis, and we have
at least a general outline of the way miRNAs modulate protein translation
and mRNA stability (discussed elsewhere in this book), we still know
relatively little about the specific biological functions of most of them.
Fortunately, this knowledge gap is rapidly being filled. Genetic experiments in mice are showing that miRNAs are key elements in regulatory
circuitries controlling a wide array of biological processes, including cell
fate determination, cell survival, proliferation and differentiation. Perhaps,
not surprisingly, with these studies came also the realisation that genetic
and epigenetic changes in miRNAs are involved in the pathogenesis of
human diseases, including cancer.

The objective of this chapter is to provide a brief overview of what we
currently know about the specific functions of individual miRNAs in
mammals. Due to space limitations, the chapter will focus on a few para-
digmatic examples that illustrate the profound impact of miRNAs on
embryonic stem cells and mammalian development.

4.2 Processing of miRNAs

The biogenesis of miRNAs begins with the transcription of a longer precursor (pri-miRNA) by RNA Polymerase II.[1] Pri-miRNAs can be derived
either from intergenic regions or from introns of protein-coding genes
(intronic miRNAs). In the nucleus, the pri-miRNA is cleaved by the
RNAse Drosha to generate a stem-and-loop precursor (pre-miRNAs)[2–4]
that is exported to the cytoplasm in an Exportin-5-dependent manner[5]
and further processed by a second RNAse (Dicer) to produce the mature
miRNA.[6] Frequently, multiple adjacent miRNAs are transcribed into a
single pri-miRNA and processed in parallel (miRNA clusters).[7] In animals,
miRNAs bind via partial sequence complementarity to the 3'-untranslated
region (3' UTR) of target mRNAs and inhibit their translation.[8–10] In
addition, miRNA can reduce the expression of their targets by negatively

affecting mRNA stability.[11,12] Computational and experimental evidence indicates that in any given miRNAs, the nucleotides at positions 2–7 (the so-called 'seed region') are the critical determinants of target specificity.[12–15]

While most microRNAs act more as rheostats than as on-off switches, producing relatively modest changes in protein output,[16,17] their overall impact on gene expression is profound and pervasive, with between 30% and 50% of human genes being predicted to be subjected to modulation by one or more miRNAs.[18,19]

4.3 Deletion of Dicer and Drosha in ES Cells

A convenient way to probe the functions of miRNAs is to disrupt their biogenesis, via genetic ablation or knockdown of Dicer or Drosha. In mice, deletion of Dicer leads to early embryonic lethality at or before embryonic day 7.5.[20] Using a conditional gene targeting strategy, two groups have generated and characterised Dicer-deficient embryonic stem cells (ES).[21,22] Despite being unable to generate miRNAs, Dicer-null ES cells are viable and retain normal ES cell characteristics, including morphology and expression of ES cell markers such as Oct4. However, their growth rate is significantly reduced compared to wild-type or Dicer+/– ES cells, and their ability to differentiate *in vitro* and *in vivo* is also severely compromised, suggesting that miRNAs may be required for terminal differentiation.

In a number of organisms, RNAi has been proposed to control heterochromatin formation, particularly at centromeres. The hypothesis is that transcripts derived from centromeric repeats form double-stranded RNAs that are processed by Dicer into siRNAs that are then used to nucleate or maintain centromeric heterochromatin. In murine ES cells, deletion of Dicer indeed leads to accumulation of centromeric transcripts, although different studies have provided contradictory results with respect to the nature of the underlying defect. While one group suggested loss of DNA methylation and changes in the pattern of histones modifications,[21] another report did not confirm these findings.[22]

The experiments discussed above strongly suggest that Dicer-dependent small RNAs are essential for early embryonic development and for cell differentiation. However, Dicer is also essential for the processing of other small RNAs, including endogenous small interfering RNAs (siRNAs) and short hairpin RNAs (shRNAs), so it remains difficult to unequivocally attribute the observed phenotype to loss of miRNAs.

Targeted deletion of DGCR8, an essential component of the microprocessor complex that catalyzes the conversion of pri-miRNAs to pre-miRNAs in the nucleus, helps to address this point. DGCR8 is in fact dispensable for the biogenesis of siRNAs, shRNAs and mirtrons, but it is absolutely required for the generation of miRNAs.[23] In many aspects, the deletion of DGCR8 resembles the deletion of Dicer. DGCR8-deficient mice arrest early in development, and DGCR8-/- ES cells proliferate more slowly than wild-type cells and spend relatively more time in the G1 phase of the cell cycle.[24] While the ability to differentiate is also compromised in DGCR8-/- ES cells, the defect is somewhat less severe than in Dicer-/- ES cells. Interestingly, the proliferation defect of DGCR8-/- ES cells has been studied in more detail and can be largely rescued by re-expressing members of the miR-290 family of ES-specific miRNAs.[24] MiRNAs belonging to this family can modulate the Cdk2-cyclinE-Rb pathway by repressing the expression of Cdkn1a, Lats2 and Rbl1 and Rbl2, although a formal demonstration that these are the critical targets has not been provided yet.

4.4 Conditional Deletion of Drosha, DGCR8 and Dicer

The availability of Dicer and DGCR8 conditional knockout alleles provide an excellent opportunity to investigate the functions of small RNAs in a variety of tissues and at various developmental stages, the only limitation being the availability of the appropriate Cre-expressing strain.

Table 4.1 summarises some of the published reports of the consequences of ablating miRNA biogenesis in a variety of tissues. Perhaps not surprisingly, in most cases, ablation of Dicer has profound effects on cell survival and/or cell differentiation. In interpreting the results of these experiments it is important to keep in mind that Cre-mediated recombination is rarely complete and thus there may be a strong selective pressure favouring cells that retain expression of the floxed allele. In addition, many miRNAs appear to be quite stable, therefore there may be a significant lag between the time Cre is expressed and the time miRNA expression falls to

Table 4.1. Consequences of conditional Dicer deletion in mice.

Tissue	Cre	Consequence	Reference(s)
Liver	Albumin-Cre	Steatosis, depletion of glycogen, increased proliferation (compensatory?). HCC by one year of age. Cells Dicer KO.	100
Reproductive tract	Amhr-2-Cre	Males fertile. Females sterile. Defect in oocyte transportation and inability of exogenous blastocysts to implant. Adenomyosis.	101–103
Sertoli cells	Mis-Cre	Aspermia. Sterility. Defective sertoli cells maturation.	104
Osteoclasts	CD11b-Cre	Mild osteopetrosis. Reduced number of osteoclasts.	105
Dorsal telenchephalon	Emx1-Cre	Hypotrophy of postnatal neocortex and early postnatal lethality. Increased neuronal apoptosis, block of differentiation.	106
Brain	Alpha-CAMKII-Cre	Microcephaly. Ataxia. Early postnatal lethality. Neuronal apoptosis in the cortex.	107
Cerebellum (postmitotic Purkinje cells)	Pcp2-Cre	Purkinje cells degeneration by 13 weeks of age. Ataxia.	108
Dopaminergic neurons	DR-1-Cre	Ataxia. Lethal between 10–12 weeks of age. Astrogliosis but no neurodegeneration.	109
Podocytes	Npsh2-Cre, Cd2ap-Cre	Glomerulosclerosis leading to kidney failure by 5 weeks of age.	110–112
Developing myocardium	Alpha-MHC-Cre	Dilated cardiomyopathy leading to heart failure and postnatal lethality.	81
Postnatal myocardium	Alpha-MHC-Cre-ER	Ventricular dilation, myocardial hyperthrophy, fibrosis. Sudden death in juvenile mice.	82
Endothelial cells	Tie2-Cre VECad-Cre-ER	Mice are viable. Reduced VEGF-induced angiogenesis. Reduced tumour angiogenesis. Reduced neoangiogenesis.	113
Skin	K14-Cre	Mice die soon after birth due to defective follicular development.	92, 92
	K14-Cre-ER	Thickening of epidermis. Ectopic proliferation in the suprabasal layer.	94

sub-physiologic levels. This latter point makes it often difficult to unequivocally determine the importance of miRNAs at a specific developmental stage.

4.5 Deletion of Argonaute in Mice

Following cleavage by Dicer, the resulting duplex RNA is loaded onto one of the Argonaute proteins in the context of the RISC complex. One of the two strands is retained by Ago and used as the guide strand to find target mRNAs, while the other strand is usually degraded. Ago proteins are therefore essential components of the RISC complex and are required for miRNA activity.[25] Mice and humans have eight Argonaute proteins, subdivided into two subfamilies: the Ago subfamily (human Ago1–4 and mouse Ago1–5) and the Piwi subfamily (HIWI1-3 and HILI in humans; MIWI1-2 and MILI in mice).[26] MiRNAs specifically associate with members of the Ago subfamily. While all four human Ago proteins associate with miRNAs, only Ago2 has been shown to possess endonucleolytic ('slicing') activity and only Ago2 is required for RNAi.[27,28] Consistent with this finding, targeted deletion of mouse Ago2 completely abrogates cleavage of perfectly complementary target sites by either miRNAs or siRNAs,[27] while miRNA-mediated translational repression is unaffected. Ago2-/-embryos die around or before embryonic day 9.5 and show severe defects in gastrulation, neural tube formation and cardiac development. Given that miRNA-mediated translational repression is still functional in the absence of Ago2, the severity of this phenotype cannot be easily explained by a global loss of miRNA function. Perhaps Ago2 is absolutely required by a subset of miRNAs that act during early mammalian development. Alternatively, loss of the 'slicing' activity of Ago2 could be responsible for the severity of the phenotype. This could be related to a few miRNAs that have nearly perfectly complementary binding sites on target mRNAs. One such miRNA (miR-196) has indeed been shown to cleave the *Hox8B* mRNA,[29] and it cannot be excluded that the true extent of miRNA-mediated mRNA cleavage in mammalians has been underestimated. The generation of mice carrying a catalytically inactive Ago2 allele will contribute to address this point, but it is worth noting that at least in the hematopoietic system, a catalytically inactive Ago2 transgene can successfully rescue the phenotype caused by Ago2 deletion (O'Carroll *et al.*, 2007),[30] suggesting that, at least in this context, miRNA-mediated mRNA cleavage does not play a critical role.

Knockout mice for the other Ago genes have not been reported yet, so the degree of functional overlap during mammalian development is unknown. However, at least in murine ES cells, the simultaneous deletion of Ago1, Ago3 and Ago4 does not compromise miRNA-mediated repression, thus confirming that Ago2 alone is capable of mediating both mRNA cleavage and translational repression.[31] Importantly, the additional deletion of Ago2 renders these cells completely refractory to miRNA-mediated gene silencing, indicating that collectively Ago1–4 are required for miRNA function in ES cells. This finding also shows that Ago5 — the other Ago member in mice (not present in the human genome) — is dispensable.

4.6 Biological Functions of Individual miRNAs

Collectively, the studies discussed in Section 4.5 provide compelling evidence for the importance of miRNAs in early mammalian development, stem cell renewal and differentiation. There are however important limitations to these kind of studies. Firstly, with a few notable exceptions, they do not allow the identification of the key miRNA(s) whose loss is responsible for the observed phenotype. Secondly, it is difficult to exclude the possibility that some of these phenotypes are actually independent from the loss of miRNA function.

The functional characterization of individual miRNAs in mammals is complicated for several reasons. Empirical and computational considerations indicate that a single miRNA can potentially modulate the expression of hundreds of different genes.[12–15,18,19] In addition, a given mRNA often has binding sites for more than one miRNA, thus exponentially increasing the combinatorial potential of miRNA-mediated regulation. Finally, the ability of a miRNA to repress its targets may also be influenced by the presence or absence of additional proteins that facilitate or antagonise the ability of the RISC complex to act on the target mRNA.[32–34] Thus the net effect of a particular miRNA will not only depend on whether and how much of it is present in a given cell, but also on additional parameters such as the expression levels of its potential targets and the overall miRNA milieu of the cell. It is therefore not surprising that although increasingly accurate prediction algorithms to identify the potential targets of miRNAs have been devised, it is not possible yet to accurately predict the function of a miRNA solely on the basis of its predicted targets.

So far, the study of mice carrying loss-of-function and gain-of-function alleles has proven to be the best approach to assign specific functions to individual miRNAs. Unfortunately, the common occurrence of multiple copies (paralogues) of the same miRNA in the mammalian genomes can complicate the execution and interpretation of such studies. For example, the mouse genome contains a dozen of *let-7* family members, with identical or nearly identical sequence, spread over multiple chromosomes.[35] Although the specific expression pattern of each *let-7* miRNA is not known yet, it appears that multiple members are commonly co-expressed in the same cell type. A consequence of this redundancy is that targeted deletion of only one member of a miRNA family may not reveal an obvious phenotype due to functional compensation by the other members. One such example is represented by the *miR-17~92* family of miRNA clusters, which consists of three highly-related polycistronic miRNA clusters (Section 4.10).

4.7 MicroRNAs in Hematopoiesis and the Immune System

Among the various cell differentiation programmes, the formation of blood cells from hematopoietic stem cells is arguably the best studied and the most amenable to genetic manipulation. It is therefore not surprising that miRNAs affecting this process were among the first to be characterised in mice. Studies with conditional knockout alleles of Dicer, Drosha and Ago2 have made it clear that miRNA function plays a critical role in the differentiation and survival of B and T lymphocytes, as well as erythrocytes. Specific ablation of Dicer in the thymus at the double negative stage (DN3) profoundly reduces the number of peripheral T cells, largely due to increased apoptosis.[36] The specific effect on T cell maturation is affected by the choice of the Cre-strain used to delete Dicer, a finding which suggests that miRNAs play different roles throughout the differentiation of T cells. For example, deletion of Dicer at the double-positive stage (using a CD4-Cre strain) leads to a relatively modest (2-fold) reduction in the number of peripheral T cells, but significantly impairs their ability to differentiate into the Th2 lineage.[37] The generation of T regulatory cells is also greatly impaired in these mice, and the lack of immune regulation eventually leads to the severe immuno-pathology and death of the mutant mice.[36,38] The effect on T regulatory cells is cell autonomous, as the specific deletion of Dicer in this subclass of T cells is sufficient to cause a severe reduction of T regulatory cells and leads to a fatal autoimmune disease that is indistinguishable from that caused by the deletion of Foxp3 itself, the master regulator of the development of T regulatory cells.[39] Analogously, in B cells a central role for miRNAs in allowing the survival and maturation of pro- and pre-B cells was first suggested by genetic deletion of Dicer and Ago2.[30,40]

4.8 MiR-181a

The discovery that miR-181 can affect lineage choice during hematopoiesis provides strong direct evidence that a single miRNA can modulate a complex differentiation pathway in mammals. Overexpression of miR-181a in hematopoietic stem cells is sufficient to push towards B cell development and to repress the development of CD8+ T cells.[41] A subsequent, more detailed characterisation of the role of miR-181a in T cell development has provided a nice example of how this miRNA can profoundly modulate the response of developing and mature T cells to antigens. By doing so, miR-181 plays a critical role in assuring that the immune system efficiently discriminates between self and non-self antigens.[42] MiR-181a expression is 10-fold higher in CD4+CD8+ double positive (DP) thymocytes than in mature T cells. *In vitro* studies using miRNA mimics and antagonists have shown that in DP cells miR-181a increases the sensitivity and strength of the T cell receptor signalling pathway. It does so at least in part by directly suppressing a group of phosphatases that act on and inactivate key signal transduction molecules.[42] Thus, different expression levels of miR-181a could explain why DP cells can respond to weak self-antigens that fail to induce activation of mature T cells. As always, overexpression experiments have to be interpreted with caution, as they can lead to off-target effects and non-physiologic responses, and it will be important to determine the consequences *in vivo* of loss of function of miR-181a. Despite these caveats, these studies provide an elegant example of how a single miRNA can function as a rheostat in modulating the output of a signalling pathway.

4.9 MiR-155

The gene encoding the primary transcript for *miR-155* had been identified well before the discovery of miRNAs, as a common proviral DNA insertion site in lymphomas induced by the avian leukosis virus.[43] The absence of an obvious open reading frame remained a puzzling feature of the *BIC* oncogene (as it was initially named) even after it was shown that it could cooperate with *Myc* (avian myelocytomatosis viral oncogene homolog) in inducing hematopoietic tumours.[44] Although the observation that the *BIC* RNA can form extensive secondary structures (including a 145-basepair stem-loop that we now know is the precursor to *miR-155*) suggested that the RNA itself could be the oncogenic factor,[45] its mechanism of action remained unclear until the identification of *miR-155*. The study of genetically-engineered mice with gain- and loss-of-function alleles of *miR-155* has provided valuable insights into its physiological and oncogenic properties. Ectopic expression of *miR-155* in the bone marrow of mice has been reported to induce either polyclonal pre-B cell proliferation followed by full blown B cell leukaemia[46] or myeloproliferation,[47] depending on the system used to drive expression of the transgene.

Although *miR-155* is dispensable for normal B and T cell development, *miR-155*-deficient mice have defective B and T cells.[48–50] In particular, these mice display a reduced germinal centre B-cell reaction and defective IgG production in response to immunisation with either T cell-dependent or independent antigens (IgM production is normal). These results suggest an important role for *miR-155* in the generation of isotype-switched, high-affinity antibodies. Among the several targets of *miR-155* that may mediate its function is the gene-encoding activation-induced cytidine deaminase (AID), which allows immunoglobulin diversification by promoting somatic hypermutation and class-switch recombination in B cells. Two groups have recently demonstrated that mutation of the single *miR-155* binding site in the 3' untranslated region (3' UTR) of the *AID* gene partially phenocopies deletion of *mir-155* itself.[51,52] In both cases, the result is increased levels of AID in germinal centre B cells and impaired affinity maturation in response to antigen stimulation, but the phenotypic overlap is only partial. In fact, while class-switch recombination is reduced in mice lacking *mir-155*,[49,50] it is increased in mice carrying mutant AID.[51,52] Thus, although these experiments elegantly demonstrate how loss of miRNA-mediated regulation of a single target gene can have profound physiological effects, they also serve as an important reminder that the phenotypic consequences of loss of a miRNA are likely due to the simultaneous deregulation of multiple target mRNAs.

4.10 *MiR-17~92* and Paralogues

MiR-17~92 occupies a special place among the oncogenic miRNAs, being the first polycistronic miRNA cluster to be associated with the pathogenesis of human cancers.[53] The first indication of a role of *mir-17~92* in tumourigenesis was the observation that this locus is frequently amplified in a subset of human B cell lymphomas known as 'diffuse large B cell lymphomas'.[53] Subsequently, overexpression or increased copy number of *mir-17~92* has been reported in other tumour types, including carcinomas of the breast, lung and colon, as well as neuroblastomas and medulloblastomas.[54–58]

The argument that *mir-17~92* is a *bona fide* oncogene is strengthened by the finding that a truncated version of this cluster (*mir-17~19b*, lacking mir-92) greatly accelerates tumourigenesis in a mouse model of B cell lymphoma in which the initiating oncogene is an Eμ-*Myc* transgene.[56] Interestingly, not only does *mir-17~92* cooperate with the *Myc* oncogene, but it is also one of its transcriptional targets.[59] A molecular circuit has emerged, linking *Myc*, *mir-17~92* and members of the E2F family of transcription factors. According to this model, c-*Myc* can directly induce transcription of *mir-17~92*, and two members of this cluster (*mir-17* and *mir-20*) can in turn repress the translation of E2F proteins (E2F1, E2F2 and E2F3). The circuitry is closed by E2F1–3 which can also induce the transcription of *mir-17~92*.[59–61]

What is the significance of these findings with respect to the oncogenic properties of *mir-17~92*? E2F proteins are key regulators of cell cycle progression, but are also inducers of apoptosis.[62] This latter function is likely mediated via activation of the p19-Arf/p53 pathway. It is tempting to speculate that the modulation of E2F1–3 expression by *miR-17~92* may contribute to *Myc*-induced transformation by suppressing cell death. This would be consistent with the observation by He *et al.* that in the Eμ-*Myc* model B cell lymphomas overexpressing *miR-17~19b* show greatly reduced spontaneous apoptosis compared to controls.[56] While repression of E2F1–3 is likely to contribute to the oncogenic activity of *miR-17~92*, it does not explain it fully. Indeed, deletion of E2F1 not only does not accelerate Eμ-*Myc*-induced lymphomagenesis, but it actually suppresses it.[63] Indeed a number of other targets of *miR-17~92* have also been proposed to mediate its tumour promoting activities, among them PTEN, p21, AML1, Bim, TSP1 and CTGF.[54,64–68]

The recent generation of loss-of-function and gain-of-function alleles of *miR-17~92* in the mouse has provided important insights into the physiologic functions of this cluster and its role in tumourigenesis.[67–69] Transgenic mice overexpressing *miR-17~92* in lymphocyte progenitors develop a lymphoproliferative disorder affecting both B and T cells that eventually results in autoimmunity.[68] At the other end of the spectrum, in mice carrying a homozygous deletion of the *miR-17~92* locus, B cells undergo premature cell death at the pro-B/pre-B stage, resulting in lymphopenia.[67]

These findings lend additional support to the hypothesis that one or more miRNAs expressed from the *miR-17~92* are critical regulators of cell survival in lymphocytes. While we are far from a full identification of the key set of genes controlled by *miR-17~92*, the works of Ventura and Xiao[67,68] suggest that the effect on cell survival may be mediated, at least in part, by the BH3-only protein Bim (BCL2L11). Both authors have reported that the Bim 3′ UTR contains multiple binding sites for miRNAs expressed by *miR-17~92* and that the expression of this pro-apoptotic gene is suppressed by this cluster. Bim is a crucial regulator of cell survival in lymphocytes[70,71] and a potent suppressor of *Myc*-induced B cell lymphomagenesis.[72] Moreover, its tumour suppressor function is dosage-sensitive, as even deletion of a single Bim allele leads to accelerated lymphomagenesis in the Eμ-*Myc* model without concomitant loss of the remaining wild-type allele.[72] It is therefore likely that Bim suppression by *miR-17~92* contributes to both the tumour promoting activity of *miR-17~92* overexpression and to its physiologic function in regulating normal B cell development. In support of this hypothesis, Koralov and colleagues have found that deletion of the miRNA processing enzyme Dicer in B cell progenitors leads to a drastic defect in B cell proliferation that is reminiscent of what was observed in *miR-17~92*-deficient mice.[40] This defect is accompanied by elevated expression of Bim and, more importantly, can be partially rescued by Bim deletion.[40]

The analysis of *miR-17~92* KO mice is also shedding some light on additional functions of this cluster.[67] *MiR-17~92*-deficient mice are significantly smaller than their wild-type littermates, have severely hypoplastic lungs, incompletely closed interventricular septum, and die within few minutes after birth. The lung hypoplasia is of particular interest, as *miR-17~92* overexpression — and occasionally amplification of the locus — has been reported in human lung cancers.[55,73] The mechanism underlying the lung hypoplasia is currently unclear, but reduced cell proliferation may play a role. Consistent with this hypothesis, forced expression of *miR-17~92* under the control of a lung-specific promoter leads to increased proliferation and blocks differentiation of lung epithelium *in vivo*, an effect that could be mediated, at least in part, by deregulation of the RBL2 gene.[69]

A recurring theme with miRNAs is the frequent occurrence of multiple copies of the same, or of closely related, miRNAs at different loci. This fact raises the possibility of functional overlap and compensation, and significantly complicates the interpretation of loss-of-function studies where only one member of a miRNA family is deleted. The *miR-17~92* cluster is no exception to this rule. Both the human and the mouse genomes contain two closely related paralogues: *miR-106a~363* (on the X chromosome), and *miR-106b~25* (on chromosome 7 in humans and chromosome 5 in mice). These two paralogues express miRNAs that are highly similar and in some cases identical to those encoded by *miR-17~92*. While greater emphasis has been placed on *miR-17~92*, emerging evidence suggests that *miR-106a~363* and *miR-106b~25* are also potential oncogenes. *MiR-106b~25* is overexpressed in gastric cancers, where it impairs signalling through the TGF-beta pathway,[66] and *miR-106a~363* has been reported to be frequently activated by insertional mutagenesis in mouse models of

T cell leukaemias and lymphomas.[74–76] To determine the extent of functional cooperation between these three highly-related miRNA clusters, Ventura *et al.* have generated KO mice for each of them, as well as compound mutant mice in which two or all three clusters were simultaneously deleted. Their analysis shows that only *miR-17~92* is essential for mouse development, as even double knockout mice simultaneously deficient for *miR-106a~363* and *miR-106b~25* develop normally and are fertile. Interestingly, while deletion of *miR-106b~25 per se* does not result in any obvious defect, simultaneous deletion of *miR-106b~25* and *miR-17~92* leads to embryonic lethality at mid-gestation, with embryos showing severe developmental defects and increased apoptosis, a phenotype much more severe than that caused by the deletion of *miR-17~92* alone.[67] At the moment it is unclear why *miR-106b~25* and *miR-106a~363* are not essential genes while *miR-17~92* is. It must be noted, however, that *miR-106a~363* is expressed at much lower levels compared to *miR-17~92* in most tissues and that the *miR-106b~25* cluster does not contain miRNAs of the miR-19 and miR-18 family (which are present in the *miR-17~92* cluster).[67] It is likely that the generation of more refined mutant mice, in which individual miRNAs belonging to the *miR-17~92* cluster are deleted or added back, will answer these important questions.

4.11 MiR-150

Mounting experimental evidence confirms the computational prediction that a single miRNA can simultaneously modulate the expression of multiple genes. In the vast majority of cases, the changes in protein output caused by the ectopic expression (or the genetic deletion) of a miRNA are however relatively modest.[16,17] The miR-150/c-Myb circuit, and its role in lymphocyte development, represents one of the best examples of how even relatively small changes in the expression level of a miRNA and its effect on a single mRNA can have profound biological consequences. MiR-150 is highly expressed in mature lymphocytes, but not in their progenitors. The expression of one of its targets, the transcription factor c-Myb follows the opposite pattern, being highest in lymphocyte progenitors and being downregulated as their differentiation proceeds. C-Myb is a well-characterised regulator of lymphocyte development and its expression is required for the transition from the pro-B to the pre-B stage.[77,78] In an elegant series of genetic experiments, using both loss-of-function and gain-of-function alleles of miR-150, Xiao *et al.* have demonstrated the importance of this miRNA in controlling lymphocyte development *in vivo.*[79] Ectopic expression of miR-150 in lymphocyte progenitors causes a dose-dependent block of differentiation at the pro-B to pre-B cell stage. Interestingly, even the highest dose of miR-150 causes only a relatively modest (35%) reduction of c-Myb expression. This effect is however enough to explain the phenotype of miR-150 transgenic mice, as a similar B cell maturation defect is seen in c-Myb heterozygous mice. Whether c-Myb is the only relevant target of miR-150 in lymphocyte development remains to be demonstrated and will likely require the analysis of mice carrying a mutation in the miR-150 binding site on c-Myb.

4.12 MiR-223

MiR-223 is another nice example of a miRNA fine-tuning blood cell development and function. MiR-223 expression is tightly controlled during haematopoiesis and seems largely restricted to the more mature stages of neutrophil differentiation. While miR-223-null mice develop normally, are fertile and don't show any obvious phenotype, their peripheral blood and their bone marrow contain an increased number of neutrophils (neutrophilia) and a shift towards neutrophil with hyper-segmented nuclei.[80] The neutrophilia is likely the result of increased proliferation and differentiation of committed progenitor cells, rather than decreased clearance of neutrophils. Interestingly, miR-223 seems to negatively control not only the number of neutrophils, but also their response to a variety of stimuli. This is demonstrated by the hyper-reactivity of miR-223-null neutrophils in response to lipopolysaccharide and phorbol myristate acetate (PMA).[80] MiR-223-null neutrophils are also more efficient in killing pathogens, such as the yeast *Candida albicans*, but this increased reactivity comes at the expense of an increased incidence of inflammatory lung pathology in older animals.

In contrast to what was reported for miR-150, where deregulation of a single target can explain the phenotype of miR-150-/- mice, it appears that miR-223 controls neutrophil development and function by modulating multiple pathways. For example, an increased expression of the miR-223 target Mef2c (a transcription factor) in miR-223-null mice is likely responsible for the neutrophilia, as this phenotype is rescued by the deletion of Mef2c in neutrophil progenitors.[80] But miR-223Mef2c compound mutant neutrophils still display hyper-segmentation and hyper-reactivity, thus suggesting the existence of at least one additional, functionally relevant target.

4.13 MiRNAs in Heart Development and Function

That miRNAs are essential for normal skeletal muscle and heart development and function was initially demonstrated by a series of studies examining the consequences of tissue-specific deletion of Dicer during embryonic as well as postnatal muscle development in mice.[81–84] Ablation of Dicer in cardiomyocytes at E8.5 results in embryonic lethality by E12.5 due to failed ventricular development.[82,84] Even when Dicer is removed in postnatal cardiomyocytes, loss of miRNAs leads to impaired cardiac function and premature death.[82] Analogously, embryonic Dicer deletion in myoblasts leads to severe muscle hypoplasia, which results in prenatal lethality.[83]

While these studies make clear that miRNAs are important players in the heart and the muscles, the identity and mechanism of action of the individual miRNAs responsible for these phenotypes is only beginning to be uncovered.[85]

4.14 The miR-1~133 Family of miRNA Clusters

The mouse genome contains three paralogue muscle-specific bicystronic clusters: miR-1-1~133a-2, miR-1-2~133a-1 and miR-206~133b. Each cluster consists of one member of the miR-1/206 family and one member of the miR-133 family. While the two families have unrelated sequences and are predicted to target largely non-overlapping sets of genes, members of the same family are identical or nearly identical (for example, miR-1-1 and miR-1-2 share the same sequence, as do miR-133a-1 and miR-133a-2). These three clusters exhibit remarkably restricted expression patterns.

MiR-1-1~133a-2 and miR-1-2~133a-1 are expressed in both cardiac and skeletal muscle, while miR-206~133b is only expressed in skeletal muscle. Cardiac expression is mediated by the binding of MEF2 and SRF transcription factors to specific cis-regulatory elements, while myoD (myoblast determination protein 1) mediates expression of these clusters in the skeletal muscle. Given their exquisitely restricted expression pattern, these miRNAs have been subject to intense investigation. Loss-of-function experiments in mice have shown that deletion of miR-1-2 leads to ventricular septal defect and lethality between late embryogenesis and birth.[84] The penetrance of this phenotype is only partial (~50% of mutant mice), and the mice that survive adulthood are normal with the exception of some minor electrophysiological anomalies. Because miR-1-1 expression is intact in these mice, the phenotype reflects only a partial loss of miR-1 function and it will be important to examine the phenotype of combined loss of miR-1-1 and miR-1-2.

MiR-133a-1 and miR-133a-2 also share the same sequence and are expressed at similar levels in the heart and the muscle. In this case, deletion of either miR-133a-1 or miR-133a-2 does not lead to any obvious phenotype, and the mice are viable and fertile. In contrast, compound mutant mice lacking both miRNAs, show a dramatic phenotype.[86] A fraction of double knockout (DKO) mice die *in utero* or soon after birth (~50%) and only ~25% of DKO mice are still alive by postnatal day 10. Not surprisingly, cardiac development is seriously affected by the loss of these miRNAs and the presence of a large ventricular septal defect leads to perinatal lethality. This defect is not present in a minority of DKO mice that reach adulthood, although they develop extensive fibrosis by 2–4 months of age, which is accompanied by reduced ventricular function and followed by dilated cardiomyopathy by 5–6 months of age. Interestingly, these mice have normal skeletal muscle development, likely due to the redundant activity of miR-133b. The exact mechanism of action of miR-133a is not known yet, but the analysis of the DKO earths revealed aberrant myocyte proliferation, increased apoptosis of myocytes of the ventricular septum, increased expression of Cyclin D2 and SRF, as well as ectopic expression of smooth muscle genes. Interestingly, the opposite phenotype is observed in transgenic mice overexpressing miR-133a in the developing heart.[86]

4.15 MyomiRs

One interesting feature of a sizeable fraction of mammalian miRNAs is that they are embedded into introns of protein-coding genes and, at least in some cases, the expression of the miRNA and the host protein-coding gene appears to be co-regulated.[87] The family of 'myomirs', so called because of their association with myosin genes, represents a good example of why such a set-up might be favoured by natural selection.[88] In the mouse genome there are three myomirs: miR-208, miR208b and miR-499. Each myomir is embedded within an intron of a myosin heavy chain (MHC) gene. MiR-208 is within *Myh6* (α-MHC), miR-208b within *Myh7* (β-MHC) and miR-499 within *Myh7b* (encoding for another β-type MHC). These three microRNAs share not only a similar genomic organisation, but are also related in their sequence sharing the same 'seed'. The ratio between α-MHC and β-MHC in the heart determines cardiac performance and these two types of myosin fibre are regulated in an antithetical manner. In the adult mouse, the α-MHC is the predominant isoform and its expression is induced by thyroid hormones, while β-MHC expression is maintained at very low levels. In contrast, in response to hypothyroidism or to pressure overload, the ratio between α- and β-MHC rapidly inverts, leading to cardiac remodelling and reduced contractility.

While the function of miR-208b and miR-499 has not been determined yet, the study of knockout mice for miR-208 has revealed a surprising role of this miRNA in the regulation of MHC expression.[89] MiR-208-/- mice develop normally and show no apparent defects in cardiac shape, indicating that this miRNA is not required for cardiac morphogenesis. The most striking consequence of the loss of miR-208 is an almost complete inability to upregulate β-MHC expression in response to hypothyroidism or pressure overload. As a consequence of this failure, the heart of miR-208-/- mice does not show the signs of hyperthrophy and fibrosis that are observed in the heart of wild-type mice subjected to the same treatment.

Thus, a microRNA encoded within the α-MHC gene is required for the upregulation of the β-MHC isoform in the adult heart in response to stress or hypothyroids (but not for the physiologic β-MHC expression during embryonic development). The molecular mechanisms underlying this phenomenon are not yet entirely understood, but one interesting target of miR-208 has been proposed. The gene encoding the thyroid hormone receptor associated protein 1 (THRAP1) is important for the ability of T3 to repress β-MHC expression. In miR-208-/- mice, THRAP1 levels are increased and this likely contributes to enhancing the repressive activity of T3 on β-MHC expression.[89]

Of course, these studies are only beginning to uncover the functions of myomirs, and we know nearly nothing about the other two members of this family: miR-208b and miR-499. It also remains unclear how much functional overlap exists between these three microRNAs and the extent to which they regulate a common set of targets.

4.16 MiR-203 and Skin Development

The skin is the largest organ of the human body and is characterised by continuous cell renewal throughout a human life.[90] Keratinocytes in the basal layer periodically exit the cell cycle and begin a carefully orchestrated differentiation process that is accompanied with the progressive migration into the outer layers (spinous layer, granular layer and stratum corneum). This process is propelled by skin stem cells that reside in the basal layer, where they constitute only 0.01% of the total population,[91] and in the 'bulge' located at the base of hair follicles.

The first indication that microRNAs are important for postnatal skin homeostasis came from the analysis of mice with conditional deletion of Dicer in the basal layer.[92,93] These mice appear normal at birth but die within one week due to severe dehydration. Postnatal skin homeostasis is severely compromised in these mice. Histologically, the hair germs, rather than invaginating into the dermis and giving origin to the hair follicles, instead evaginate into the epidermis and form cysts. The cysts – that contain cell types found in the normal hair follicles – eventually enlarge and disrupt epidermal integrity, compromising the barrier function of the skin and thus leading to the postnatal lethality. In addition, genetic ablation of Dicer in the epidermis of adult mice using a K14-Cre-ER transgene has allowed Yi *et al.* to uncover a crucial role of miRNAs in repressing the proliferation of suprabasal cells.[94] Combined with the observation of a very similar phenotype in mice subjected to conditional ablation of DGCR8, these results strongly indicate a critical role for miRNAs in skin morphogenesis.[95]

With respect to the identity of the miRNA(s) mediating these phenotypes, convincing evidence points to a crucial role for miR-203. MiR-203 expression is undetectable in basal cells, but is rapidly and potently induced in suprabasal cells in coincidence with the exit from the cell cycle.[94] In transgenic mice ectopically expressing miR-203 in the basal layer of the epidermis, by postnatal day 0 (P0) the skin is reduced to a single layer of basal cells and a single layer of suprabasal cells. In addition, as development proceeds, the cells in the basal layer progressively lose expression of basal progenitor/stem cell marker Keratin 5 (K5) and are replaced by cells expressing the suprabasal marker K10.[94] Consistent with a crucial role for miR-203 in restricting the proliferative potential of suprabasal cells, antagonising miR-203 function results in an expansion of cells expressing basal cell markers and ectopic proliferation in the suprabasal layers. Collectively these results demonstrate that induction of miR-203 expression is required to restrict the proliferation potential of basal cells as they transition to the suprabasal layer. Intriguingly, the skin phenotype observed in miR-203 transgenic mice is very similar to that observed in p63-/- animals.[96,97] In addition, expression of ΔN-p63α (the main p63 isoform in the epidermis) inversely correlates with expression of miR-203, being abundant in the basal layer and rapidly downregulated as the cells exit the cell cycle and begin their migration into the suprabasal compartment. Even more intriguingly, ectopic expression of miR-203 is sufficient to downregulate the expression of ΔN-p63α via direct binding to two sites in its 3' UTR.[94,98] Taken together, these findings point to a model in which induction of miR-203 leads to the rapid downregulation of ΔN-p63α, which in turn is required to allow exit from the cell cycle and terminal differentiation of epidermal cells.

Despite the elegance of this model, several questions remain to be addressed. In particular, the generation of mice expressing ΔN-p63α with a mutated 3' UTR will be required to confirm the relevance of ΔN-p63α as a physiologic target of miR-203. In addition, it will be interesting to determine whether miR-203 plays a similar role in suppressing 'stemness' in other stratified epithelia and in other tissues, and to identify the mechanism through which miR-203 expression is so tightly controlled.

Finally, the role of miR-203 in restricting the proliferative potential of progenitor cells suggests the possibility that its deletion may promote tumourigenesis in the skin or in other organs. Indeed, recurrent genetic deletion or epigenetic silencing of miR-203 has been recently reported in chronic myelogenous leukaemias.[99]

4.17 Conclusion

This brief review chapter provides only a glimpse into the complex and rapidly expanding field of mammalian miRNA biology. The task ahead is daunting and a full understanding of the role of miRNAs in development and in disease will require the convergence of more refined computational methods and the generation of sophisticated mouse models. Yet, it is encouraging to see how rapid our progress has been in recent years. In all tissues and systems examined so far, ablation of miRNA-mediated regulation has uncovered essential roles for this family of tiny non-coding RNAs. In some instances we are also beginning to dissect the molecular circuitries that individual miRNAs modulate under physiologic conditions and in diseases. It is safe to assume that in the near future this knowledge will increase and may even lead to novel therapeutic strategies. Whatever the ultimate outcome, it is certain that our labs, our post-docs and our students will have plenty of exciting experiments and unpredictable discoveries to make.

Acknowledgments

Work in the Ventura lab is supported by the Sidney Kimmel Foundation for Cancer Research and by the Geoffrey Beene Cancer Research Foundation.

References

1. Lee Y., Kim M., Han J. *et al.* (2004). MicroRNA genes are transcribed by RNA polymerase II. *EMBO J* **23**, 4051–4060.
2. Denli A.M., Tops B.B., Plasterk R.H. *et al.* (2004). Processing of primary microRNAs by the Microprocessor complex. *Nature* **432**, 231–235.

3. Gregory R.I., Yan K.P., Amuthan G. *et al.* (2004). The Microprocessor complex mediates the genesis of microRNAs. *Nature* **432**, 235–240.
4. Lee Y., Ahn C., Han J. *et al.* (2003). The nuclear RNase III Drosha initiates microRNA processing. *Nature* **425**, 415–419.
5. Lund E., Guttinger S., Calado A. *et al.* (2004). Nuclear export of microRNA precursors. *Science* **303**, 95–98.
6. Bernstein E., Caudy A.A., Hammond S.M. *et al.* (2001). Role for a bidentate ribonuclease in the initiation step of RNA interference. *Nature* **409**, 363–366.
7. Altuvia Y., Landgraf P., Lithwick G. *et al.* (2005). Clustering and conservation patterns of human microRNAs. *Nucleic Acids Res* **33**, 2697–2706.
8. Lee R.C., Feinbaum R.L. and Ambros V. (1993). The *C. elegans* heterochronic gene lin-4 encodes small RNAs with antisense complementarity to lin-14. *Cell* **75**, 843–854.
9. Wightman B., Ha I. and Ruvkun G. (1993). Posttranscriptional regulation of the heterochronic gene lin-14 by lin-4 mediates temporal pattern formation in C. elegans. *Cell* **75**, 855–862.
10. Filipowicz W., Bhattacharyya S.N. and Sonenberg, N. (2008). Mechanisms of post-transcriptional regulation by microRNAs: Are the answers in sight? *Nat Rev Genet* **9**, 102–114.
11. Bagga S., Bracht J., Hunter S. *et al.* (2005). Regulation by let-7 and lin-4 miRNAs results in target mRNA degradation. *Cell* **122**, 553–563.
12. Lim L.P., Lau N., Garrett-Engele P. *et al.* (2005). Microarray analysis shows that some microRNAs downregulate large numbers of target mRNAs. *Nature* **433**, 769–773.
13. Grimson A., Farh K.K., Johnston W.K. *et al.* (2007). MicroRNA targeting specificity in mammals: Determinants beyond seed pairing. *Mol Cell* **27**, 91–105.
14. Lewis B.P., Burge C.B. and Bartel D.P. (2005). Conserved seed pairing, often flanked by adenosines, indicates that thousands of human genes are microRNA targets. *Cell* **120**, 15–20.
15. Lewis B.P., Shih I.H., Jones-Rhoades M.W. *et al.* (2003). Prediction of mammalian microRNA targets. *Cell* **115**, 787–798.
16. Baek D., Villén J., Shin C. *et al.* (2008). The impact of microRNAs on protein output. *Nature* **455**, 64–71.
17. Selbach M., Schwanhausser B., Thierfelder N. *et al.* (2008). Widespread changes in protein synthesis induced by microRNAs. *Nature* **455**, 58–63.
18. Bartel D.P. (2009). MicroRNAs: Target recognition and regulatory functions. *Cell* **136**, 215–233.

19. Farh K.K., Grimson A., Jan C. *et al.* (2005). The widespread impact of mammalian microRNAs on mRNA repression and evolution. *Science* **310**, 1817–1821.

20. Bernstein E., Kim S.Y., Carmell M.A. *et al.* (2003). Dicer is essential for mouse development. *Nat Genet* **35**, 215–217.

21. Kanellopoulou C., Muljo S.A., Kung A.L. *et al.* (2005). Dicer-deficient mouse embryonic stem cells are defective in differentiation and centromeric silencing. *Genes Dev* **19**, 489–501.

22. Murchison E.P., Partridge J.F., Tam O.H. *et al.* (2005). Characterization of Dicer-deficient murine embryonic stem cells. *Proc Natl Acad Sci USA* **102**, 12135–12140.

23. Babiarz J.E., Ruby J.G., Wang Y. *et al.* (2008). Mouse ES cells express endogenous shRNAs, siRNAs, and other Microprocessor-independent, Dicer-dependent small RNAs. *Genes Dev* **22**, 2773–2785.

24. Wang Y., Baskerville S., Shenoy A. *et al.* (2008). Embryonic stem cell-specific microRNAs regulate the G1-S transition and promote rapid proliferation. *Nat Genet* **40**, 1478–1483.

25. Kim V.N., Han J., and Siomi M.C. (2009). Biogenesis of small RNAs in animals. *Nat Rev Mol Cell Biol* **10**, 126–139.

26. Peters L., and Meister G. (2007). Argonaute proteins: Mediators of RNA silencing. *Mol Cell* **26**, 611–623.

27. Liu J., Carmell M.A., Rivas F.V. *et al.* (2004). Argonaute2 is the catalytic engine of mammalian RNAi. *Science* **305**, 1437–1441.

28. Meister G., Landthaler M., Patkaniowska A. *et al.* (2004). Human Argonaute2 mediates RNA cleavage targeted by miRNAs and siRNAs. *Mol Cell* **15**, 185–197.

29. Yekta S., Shih I.H. and Bartel D.P. (2004). MicroRNA-directed cleavage of HOXB8 mRNA. *Science* **304**, 594–596.

30. O'Carroll D., Mecklenbrauker I., Das P.P. *et al.* (2007). A Slicer-independent role for Argonaute 2 in hematopoiesis and the microRNA pathway. *Genes Dev* **21**, 1999–2004.

31. Su H., Trombly M.I., Chen J. *et al.* (2009). Essential and overlapping functions for mammalian Argonautes in microRNA silencing. *Genes Dev* **23**, 304–317.

32. Bhattacharyya S.N., Habermacher R., Martine U. *et al.* (2006). Relief of microRNA-mediated translational repression in human cells subjected to stress. *Cell* **125**, 1111–1124.

33. Kedde M., Strasser M.J., Boldajipour B. *et al.* (2007). RNA-binding protein Dnd1 inhibits microRNA access to target mRNA. *Cell* **131**, 1273–1286.

34. Weinmann L., Hock J., Ivacevic T. *et al.* (2009). Importin 8 is a gene silencing factor that targets argonaute proteins to distinct mRNAs. *Cell* **136**, 496–507.

35. Roush S. and Slack F.J. (2008). The let-7 family of microRNAs. *Trends Cell Biol* **18**, 505–516.

36. Cobb B.S., Nesterova T.B., Thompson E. *et al.* (2005). T cell lineage choice and differentiation in the absence of the RNase III enzyme Dicer. *J Exp Med* **201**, 1367–1373.

37. Muljo S.A., Ansel K.M., Kanellopoulou C. *et al.* (2005). Aberrant T cell differentiation in the absence of Dicer. *J Exp Med* **202**, 261–269.

38. Cobb B.S., Hertweck A., Smith J. *et al.* (2006). A role for Dicer in immune regulation. *J Exp Med* **203**, 2519–2527.

39. Liston A., Lu L.F., O'Carroll D. *et al.* (2008). Dicer-dependent microRNA pathway safeguards regulatory T cell function. *J Exp Med* **205**, 1993–2004.

40. Koralov S.B., Muljo S.A., Galler G.R. *et al.* (2008). Dicer ablation affects antibody diversity and cell survival in the B lymphocyte lineage. *Cell* **132**, 860–874.

41. Chen C.Z., Li L., Lodish H.F. *et al.* (2004). MicroRNAs modulate hematopoietic lineage differentiation. *Science* **303**, 83–86.

42. Li Q.J., Chau J., Ebert P.J. *et al.* (2007). miR-181a is an intrinsic modulator of T cell sensitivity and selection. *Cell* **129**, 147–161.

43. Clurman B.E. and Hayward W.S. (1989). Multiple proto-oncogene activations in avian leukosis virus-induced lymphomas: Evidence for stage-specific events. *Mol Cell Biol* **9**, 2657–2664.

44. Tam W., Hughes S.H., Hayward W.S. *et al.* (2002). Avian bic, a gene isolated from a common retroviral site in avian leukosis virus-induced lymphomas that encodes a noncoding RNA, cooperates with c-myc in lymphomagenesis and erythroleukemogenesis. *J Virol* **76**, 4275–4286.

45. Tam W., Ben-Yehuda D. and Hayward W.S. (1997). bic, a novel gene activated by proviral insertions in avian leukosis virus-induced lymphomas, is likely to function through its noncoding RNA. *Mol Cell Biol* **17**, 1490–1502.

46. Costinean S., Zanesi N., Pekarsky Y. *et al.* (2006). Pre-B cell proliferation and lymphoblastic leukemia/high-grade lymphoma in E(mu)-miR155 transgenic mice. *Proc Natl Acad Sci USA* **103**, 7024–7029.

47. O'Connell R.M., Rao D.S., Chaudhuri A.A. *et al.* (2008). Sustained expression of microRNA-155 in hematopoietic stem cells causes a myeloproliferative disorder. *J Exp Med* **205**, 585–594.

48. Rodriguez A., Vigorito E., Clare S. *et al.* (2007). Requirement of bic/microRNA-155 for normal immune function. *Science* **316**, 608–611.

49. Thai T.H., Calado D.P., Casola S. *et al.* (2007). Regulation of the germinal center response by microRNA-155. *Science* **316**, 604–608.
50. Vigorito E., Perks K.L., Abreu-Goodger C. *et al.* (2007). MicroRNA-155 regulates the generation of immunoglobulin class-switched plasma cells. *Immunity* **27**, 847–859.
51. Dorsett Y., McBride K.M., Jankovic M. *et al.* (2008). MicroRNA-155 suppresses activation-induced cytidine deaminase-mediated Myc-Igh translocation. *Immunity* **28**, 630–638.
52. Teng G. and Papavasiliou F.N. (2007). Immunoglobulin somatic hypermutation. *Annu Rev Genet* **41**, 107–120.
53. Ota A., Tagawa H., Karnan S. *et al.* (2004). Identification and characterization of a novel gene, C13orf25, as a target for 13q31-q32 amplification in malignant lymphoma. *Cancer Res* **64**, 3087–3095.
54. Fontana L., Fiori M.E., Albini S. *et al.* (2008). Antagomir-17-5p abolishes the growth of therapy-resistant neuroblastoma through p21 and BIM. *PLoS ONE* **3**, e2236.
55. Hayashita Y., Osada H., Tatematsu Y. *et al.* (2005). A polycistronic microRNA cluster, miR-17-92, is overexpressed in human lung cancers and enhances cell proliferation. *Cancer Res* **65**, 9628–9632.
56. He L., Thomson J.M., Hemann M.T. *et al.* (2005). A microRNA polycistron as a potential human oncogene. *Nature* **435**, 828–833.
57. Tagawa H. and Seto M. (2005). A microRNA cluster as a target of genomic amplification in malignant lymphoma. *Leukemia* **19**, 2013–2016.
58. Uziel T., Karginov F.V., Xie S. *et al.* (2009). The miR-17~92 cluster collaborates with the Sonic Hedgehog pathway in medulloblastoma. *Proc Natl Acad Sci USA* **106**, 2812–2817.
59. O'Donnell K.A., Wentzel E.A., Zeller K.I. *et al.* (2005). c-Myc-regulated microRNAs modulate E2F1 expression. *Nature* **435**, 839–843.
60. Sylvestre Y., De Guire V., Querido E. *et al.* (2007). An E2F/miR-20a autoregulatory feedback loop. *J Biol Chem* **282**, 2135–2143.
61. Woods K., Thomson J.M. and Hammond S.M. (2007). Direct regulation of an oncogenic micro-RNA cluster by E2F transcription factors. *J Biol Chem* **282**, 2130–2134.
62. Iaquinta P.J. and Lees J.A. (2007). Life and death decisions by the E2F transcription factors. *Curr Opin Cell Biol* **19**, 649–657.
63. Baudino T.A., Maclean K.H., Brennan J. *et al.* (2003). Myc-mediated proliferation and lymphomagenesis, but not apoptosis, are compromised by E2f1 loss. *Mol Cell* **11**, 905–914.

64. Dews M., Homayouni A., Yu D. *et al.* (2006). Augmentation of tumor angiogenesis by a Myc-activated microRNA cluster. *Nat Genet* **38**, 1060–1065.

65. Fontana L., Pelosi E., Greco P. *et al.* (2007). MicroRNAs 17-5p-20a-106a control monocytopoiesis through AML1 targeting and M-CSF receptor upregulation. *Nat Cell Biol* **9**, 775–787.

66. Petrocca F., Visone R., Onelli M.R. *et al.* (2008). E2F1-regulated microRNAs impair TGFbeta-dependent cell-cycle arrest and apoptosis in gastric cancer. *Cancer Cell* **13**, 272–286.

67. Ventura A., Young A.G., Winslow M.M. *et al.* (2008). Targeted deletion reveals essential and overlapping functions of the miR-17 through 92 family of miRNA clusters. *Cell* **132**, 875–886.

68. Xiao C., Srinivasan L., Calado D.P. *et al.* (2008). Lymphoproliferative disease and autoimmunity in mice with increased miR-17-92 expression in lymphocytes. *Nat Immunol* **9**, 405–414.

69. Lu Y., Thomson J.M., Wong H.Y. *et al.* (2007). Transgenic over-expression of the microRNA miR-17-92 cluster promotes proliferation and inhibits differentiation of lung epithelial progenitor cells. *Dev Biol* **310**, 442–453.

70. Bouillet P., Metcalf D., Huang D.C. *et al.* (1999). Proapoptotic Bcl-2 relative Bim required for certain apoptotic responses, leukocyte homeostasis, and to preclude autoimmunity. *Science* **286**, 1735–1738.

71. Bouillet P., Purton J.F., Godfrey D.I. *et al.* (2002). BH3-only Bcl-2 family member Bim is required for apoptosis of autoreactive thymocytes. *Nature* **415**, 922–926.

72. Egle A., Harris A.W., Bouillet P. *et al.* (2004). Bim is a suppressor of Myc-induced mouse B cell leukemia. *Proc Natl Acad Sci USA* **101**, 6164–6169.

73. Matsubara H., Takeuchi T., Nishikawa E. *et al.* (2007). Apoptosis induction by antisense oligonucleotides against miR-17-5p and miR-20a in lung cancers overexpressing miR-17-92. *Oncogene* **26**, 6099–6105.

74. Berns K., Hijmans E.M., Mullenders J. *et al.* (2004). A large-scale RNAi screen in human cells identifies new components of the p53 pathway. *Nature* **428**, 431–437.

75. Landais S., Landry S., Legault P. *et al.* (2007). Oncogenic potential of the miR-106-363 cluster and its implication in human T-cell leukemia. *Cancer Res* **67**, 5699–5707.

76. Lum A.M., Wang B.B., Li L. *et al.* (2007). Retroviral activation of the mir-106a microRNA cistron in T lymphoma. *Retrovirology* **4**, 5.

77. Bender T.P., Kremer C.S., Kraus M. *et al.* (2004). Critical functions for c-Myb at three checkpoints during thymocyte development. *Nat Immunol* **5**, 721–729.

78. Thomas M.D., Kremer C.S., Ravichandran K.S. *et al.* (2005). c-Myb is critical for B cell development and maintenance of follicular B cells. *Immunity* **23**, 275–286.

79. Xiao C., Calado D.P., Galler G. *et al.* (2007). MiR-150 controls B cell differentiation by targeting the transcription factor c-Myb. *Cell* **131**, 146–159.

80. Johnnidis J.B., Harris M.H., Wheeler R.T. *et al.* (2008). Regulation of progenitor cell proliferation and granulocyte function by microRNA-223. *Nature* **451**, 1125–1129.

81. Chen J.F., Murchison E.P., Tang R. *et al.* (2008). Targeted deletion of Dicer in the heart leads to dilated cardiomyopathy and heart failure. *Proc Natl Acad Sci USA* **105**, 2111–2116.

82. da Costa Martins P.A., Bourajjaj M., Gladka M. *et al.* (2008). Conditional dicer gene deletion in the postnatal myocardium provokes spontaneous cardiac remodeling. *Circulation* **118**, 1567–1576.

83. O'Rourke J.R., Georges S.A., Seay H.R. *et al.* (2007). Essential role for Dicer during skeletal muscle development. *Dev Biol* **311**, 359–368.

84. Zhao Y., Ransom J.F., Li A. *et al.* (2007). Dysregulation of cardiogenesis, cardiac conduction, and cell cycle in mice lacking miRNA-1-2. *Cell* **129**, 303–317.

85. Williams A.H., Liu N., van Rooij E. *et al.* (2009). MicroRNA control of muscle development and disease. *Curr Opin Cell Biol* **21**, 461–469.

86. Liu N., Bezprozvannaya S., Williams A.H. *et al.* (2008). microRNA-133a regulates cardiomyocyte proliferation and suppresses smooth muscle gene expression in the heart. *Genes Dev* **22**, 3242–3254.

87. Baskerville S. and Bartel D.P. (2005). Microarray profiling of microRNAs reveals frequent coexpression with neighboring miRNAs and host genes. *RNA* **11**, 241–247.

88. van Rooij E., Liu N. and Olson E.N. (2008). MicroRNAs flex their muscles. *Trends Genet* **24**, 159–166.

89. van Rooij E., Sutherland L.B., Qi X. *et al.* (2007). Control of stress-dependent cardiac growth and gene expression by a microRNA. *Science* **316**, 575–579.

90. Blanpain C. and Fuchs E. (2009). Epidermal homeostasis: A balancing act of stem cells in the skin. *Nat Rev Mol Cell Biol* **10**, 207–217.

91. Schneider T.E., Barland C., Alex A.M. *et al.* (2003). Measuring stem cell frequency in epidermis: A quantitative *in vivo* functional assay for long-term repopulating cells. *Proc Natl Acad Sci USA* **100**, 11412–11417.

92. Andl T., Murchison E.P., Liu F. *et al.* (2006). The miRNA-processing enzyme dicer is essential for the morphogenesis and maintenance of hair follicles. *Curr Biol* **16**, 1041–1049.

93. Yi R., O'Carroll D., Pasolli H.A. *et al.* (2006). Morphogenesis in skin is governed by discrete sets of differentially expressed microRNAs. *Nat Genet* **38**, 356–362.

94. Yi R., Poy M.N., Stoffel M. *et al.* (2008). A skin microRNA promotes differentiation by repressing 'stemness'. *Nature* **452**, 225–229.

95. Yi R., Pasolli H.A., Landthaler M. *et al.* (2009). DGCR8-dependent microRNA biogenesis is essential for skin development. *Proc Natl Acad Sci USA* **106**, 498–502.

96. Senoo M., Pinto F., Crum C.P. *et al.* (2007). p63 is essential for the proliferative potential of stem cells in stratified epithelia. *Cell* **129**, 523–536.

97. Yang A., Schweitzer R., Sun D. *et al.* (1999). p63 is essential for regenerative proliferation in limb, craniofacial and epithelial development. *Nature* **398**, 714–718.

98. Lena A.M., Shalom-Feuerstein R., Rivetti di Val Cervo P. *et al.* (2008). miR-203 represses 'stemness' by repressing DeltaNp63. *Cell Death Differ* **15**, 1187–1195.

99. Bueno M.J., Perez de Castro I., Gomez de Cedron M. *et al.* (2008). Genetic and epigenetic silencing of microRNA-203 enhances ABL1 and BCR-ABL1 oncogene expression. *Cancer Cell* **13**, 496–506.

100. Sekine S., Ogawa R., Ito R. *et al.* (2009). Disruption of Dicer1 induces dysregulated fetal gene expression and promotes hepatocarcinogenesis. *Gastroenterology* **136**, 2304–2315; e2301–2304.

101. Gonzalez G. and Behringer R.R. (2009). Dicer is required for female reproductive tract development and fertility in the mouse. *Mol Reprod Dev* **76**, 678–688.

102. Hong X., Luense L.J., McGinnis L.K. *et al.* (2008). Dicer1 is essential for female fertility and normal development of the female reproductive system. *Endocrinology* **149**, 6207–6212.

103. Nagaraja A.K., Andreu-Vieyra C., Franco H.L. *et al.* (2008). Deletion of Dicer in somatic cells of the female reproductive tract causes sterility. *Mol Endocrinol* **22**, 2336–2352.

104. Papaioannou M.D., Pitetti J.L., Ro S. *et al.* (2009). Sertoli cell Dicer is essential for spermatogenesis in mice. *Dev Biol* **326**, 250–259.

105. Sugatani T. and Hruska K.A. (2007). MicroRNA-223 is a key factor in osteoclast differentiation. *J Cell Biochem* **101**, 996–999.

106. De Pietri Tonelli D., Pulvers J.N., Haffner C. *et al.* (2008). miRNAs are essential for survival and differentiation of newborn neurons but not for expansion of neural progenitors during early neurogenesis in the mouse embryonic neocortex. *Development* **135**, 3911–3921.

107. Davis T.H., Cuellar T.L., Koch S.M. *et al.* (2008). Conditional loss of Dicer disrupts cellular and tissue morphogenesis in the cortex and hippocampus. *J Neurosci* **28**, 4322–4330.

108. Schaefer A., O'Carroll D., Tan C.L. *et al.* (2007). Cerebellar neurodegeneration in the absence of microRNAs. *J Exp Med* **204**, 1553–1558.

109. Cuellar T.L., Davis T.H., Nelson P.T. *et al.* (2008). Dicer loss in striatal neurons produces behavioral and neuroanatomical phenotypes in the absence of neurodegeneration. *Proc Natl Acad Sci USA* **105**, 5614–5619.

110. Ho J., Ng K.H., Rosen S. *et al.* (2008). Podocyte-specific loss of functional microRNAs leads to rapid glomerular and tubular injury. *J Am Soc Nephrol* **19**, 2069–2075.

111. Harvey S.J., Jarad G., Cunningham J. *et al.* (2008). Podocyte-specific deletion of dicer alters cytoskeletal dynamics and causes glomerular disease. *J Am Soc Nephrol* **19**, 2150–2158.

112. Shi S., Yu L., Chiu C. *et al.* (2008). Podocyte-selective deletion of dicer induces proteinuria and glomerulosclerosis. *J Am Soc Nephrol* **19**, 2159–2169.

113. Suárez Y., Fernández-Hernando C., Yu J. *et al.* (2008). Dicer-dependent endothelial microRNAs are necessary for postnatal angiogenesis. *Proc Natl Acad Sci USA* **105**, 14082–14087.

5

MicroRNAs in Hematopoietic Development

*Silvia Monticelli**

Innate and acquired immune responses must be tightly regulated by elaborate mechanisms to control their onset and termination. Aberrant regulation of the immune system leads to various immune-related pathological disorders, such as autoimmune diseases and leukemia. MicroRNAs (miRNAs) constitute a large family of short, non-coding RNAs that regulate gene expression in metazoans post-transcriptionally. MiRNAs are emerging as key elements in the development and functions of mammalian immune cells, as well as in determining the outcome of immune responses to infection and in the development of diseases of immunological origin. This chapter will describe advances in the knowledge of immunological relevant miRNAs, as well as new roles of miRNA biogenesis in the regulation of inflammatory diseases and in the homeostasis of the immune system.

5.1 Introduction

The hematopoietic system is a paradigm for the differentiation of distinct cell lineages from multipotent progenitors, and differentiation is modulated

*Institute for Research in Biomedicine, Via Vincenzo Vela 6, Bellinzona CH-6500, Switzerland. E-mail: silvia.monticelli@irb.unisi.ch.

by a molecular network that regulates the commitment, proliferation, apoptosis, and maturation of hematopoietic progenitor cells. Appropriate regulation of immune homeostasis is critical for ensuring adequate immunity towards harmful pathogens without uncontrolled inflammation that could lead to tissue damage. MicroRNAs (miRNAs) are small, single-stranded, non-coding RNA molecules (~22 nucleotides in length) that are emerging as major players in the regulation of endogenous gene expression in immune homeostasis. Specifically, miRNAs function mostly as endogenous translational repressors of protein-coding genes by binding to target sites in the 3′ untranslated region (3′ UTR) of messenger RNAs. Like transcription factors, miRNAs are important determinants of cellular fate specification, as well as regulators of cell proliferation and functions. Their regulatory role is underlined by the fact that deregulated miRNA expression can lead to tumorigenic transformation of cells (reviewed in Ref. 1).

Analysis of miRNA expression in cells of the immune system has shown that some miRNAs display restricted expression patterns with tissue and temporal specificities, while others demonstrate ubiquitous or constitutive expression[2-5] (see Figure 5.1). Combined analysis of miRNA expression in hematopoietic lineages, using microarray profiling along with the characterization of individual miRNAs by cloning, northern blotting and real-time quantitative RT-PCR, has resulted in the identification of miRNAs that are now considered as 'markers' of these lineages.[3-6] MiRNA profiling studies, ectopic expression, conditional genetic ablation and transgene overexpression have been critical in the investigation of the role of miRNAs in the immune system. This review will focus on the role of miRNAs in regulating differentiation and functions of cells of the immune system, and will discuss recent developments in this field as well as new directions.

5.2 MiRNA Biogenesis and Post-Transcriptional Regulation

MiRNAs are transcribed as long primary transcripts (pri-miRNA) that are mostly RNA Polymerase II-dependent and are polyadenylated.[7] The pri-miRNA contains one or more stem-loop structures, which are processed by the Microprocessor enzyme complex, comprising the ribonuclease III Drosha and the double-stranded RNA (dsRNA) binding protein

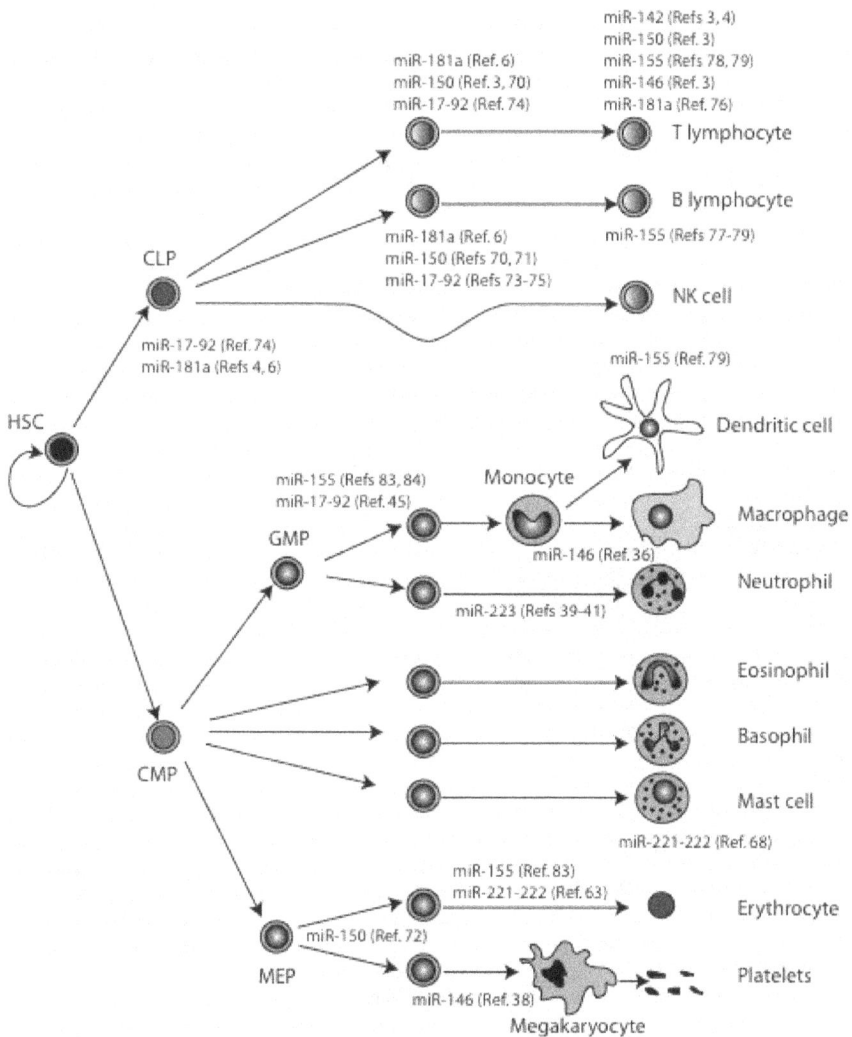

Figure 5.1. Involvement of miRNAs in immune cells differentiation and function. Schematic representation of hematopoietic lineages and miRNAs that modulate differentiation and/or function of each of the cell types. HSC, hematopoietic stem cell; CLP, common lymphoid progenitor; CMP, common myeloid progenitor; MEP, megakaryocyte-erythrocyte progenitor; GMP, granulocyte-monocyte progenitor.

Pasha/DGCR8.[8,9] This processing releases a 60- to 110-nt pre-miRNA hairpin precursor, which is further processed in the cytoplasm by Dicer into the mature miRNAs that are responsible for the post-transcriptional regulation of expression of multiple mRNAs and, hence, proteins. Specifically, the mature miRNA is loaded onto the RNA-induced silencing complex (RISC), where it associates with an Argonaute protein. The miRNA-Argounaute complex interacts with the 3′ UTR of target mRNAs through partially complementary binding of the miRNA to the target. This binding induces mainly a block in protein translation, with or without target mRNA destabilization and degradation.[10]

The maturation process of miRNAs can potentially be regulated at many steps; indeed, post-transcriptional control of miRNA expression has been reported to occur in a tissue-specific[11] and developmentally regulated fashion.[12] MiRNA precursors (either pri- or pre-miRNAs) can accumulate at high levels, but their processing can be subsequently regulated at the level of Drosha or Dicer, respectively.[11,12] The mechanism for this post-transcriptional block in miRNA processing has so far remained elusive, even though recent data have shown that Lin28, a developmentally regulated RNA binding protein, is necessary and sufficient for blockade of pri-miRNA processing of *let-7* family members during mouse embryonic stem (ES) cell differentiation.[13–15] Furthermore, TGF-β and BMP (bone morphogenetic protein) signaling have been shown to promote a rapid increase in mature miR-21 expression in human vascular smooth muscle cells through a post-transcriptional step, promoting the processing of primary transcripts of miR-21 (pri-miR-21) into precursors pre-miR-21 by the Drosha complex.[16]

MiRNAs themselves can also undergo post-transcriptional modifications. ADAR (adenosine deaminase acting on RNA) enzymes can catalyze adenosine-to inosine (A-to-I) transitions in dsRNA substrates, a process called RNA editing.[17–20] Since inosine acts like guanosine and preferentially basepairs with cytidine, editing can alter basepairing specificity. Editing of a miRNA within the region essential for target recognition (the 'seed' region, see below) can redirect the miRNA to a new set of targets, as it has been demonstrated for miR-376a.[18] Furthermore, Yang *et al.* demonstrated that editing of pri-miR-142 can inhibit Drosha cleavage, raising the possibility that editing can also alter processing of some miRNAs.[20]

Finally, miRNA-loaded RISCs can be inhibited by the RNA-binding protein Dead end 1 (Dnd1), which binds uridine-rich regions in the proximity of miRNA target sites and counteracts mRNA repression.[21] Overall, these data demonstrate that the synthesis and the target specificity of miRNAs can be modulated at several levels, therefore expanding the potential of a single genomic locus to generate multiple miRNAs with different targets.

5.3 MiRNA Target Recognition and 3′ UTR Regulation

MiRNAs regulate the expression of target genes by forming imperfect basepair interactions with target sites located in the 3′ UTRs of their target mRNAs. MiRNA target sites comprise a 'core sequence' that forms usually perfect basepairs with 7–8 bases in the 5′ end of a miRNA (the 'seed' region).[22,23] It is estimated that more than 30% of human genes are regulated by miRNAs.[22,24]

Even though the exact mechanism by which miRNAs can inhibit translation is still not completely deciphered, the end result of miRNA-mediated gene regulation is clearly a reduction in the total amount of target protein that is produced. Identifying functionally important miRNA targets is crucial for understanding miRNA functions. This was nicely shown in two recent papers demonstrating that thousands of proteins are affected by miRNAs.[25,26] Baek *et al.* used quantitative mass spectrometry to measure the response of thousands of protein in HeLa cells overexpressing miR-124, miR-1 or miR-181, as well as in mouse neutrophils deleted for *miR-223*.[25] The identities of the responsive proteins indicated that targeting is primarily through seed-matched sites in the 3′ UTR of the target mRNA. Hundreds of genes were directly repressed, although each to a modest degree. While for the targets that were repressed at highest levels mRNA destabilization usually comprised the major component of repression, the proteins from the least abundant mRNAs appeared to respond without detectable mRNA changes. In a different approach, Selbach *et al.* used a new variant of SILAC (stable isotope labeling with amino acids in cell culture) in order to measure changes in protein production between two samples.[26] In this pulsed-SILAC (pSILAC) method, cells in the two samples were labeled with two different heavy versions of amino acids.

This analysis was performed on HeLa cells transfected to individually overexpress miR-16, miR-30, miR-1, miR-155 or let-7b, or to knockdown let-7b with a locked nucleic acid (LNA) approach. Again, pSILAC revealed that miRNA overexpression had mild effects on the synthesis of most of the ~3,000 proteins quantified in each transfection. When the cellular response to let-7b overexpression and knockdown were compared, a marked anti-correlation was observed, indicating that upregulation and downregulation of let-7b levels has largely complementary effects on the proteome, and that let-7b levels can tune protein production from thousands of genes. Intriguingly, Dicer (which has several *let-7* 3'UTR seed sequence complementary sites) was one of the most strongly upregulated genes in the let-7b knockdown pSILAC. The impact of miRNA expression on the proteome shown in these two seminal works indicated that for most interactions, miRNAs act as rheostat to make fine-scale adjustments to protein output.

Interestingly, recent data have shown that animal miRNAs can exercise their control on mRNAs through targets that can reside beyond the 3' UTR.[27] Tay *et al.* identified miR-296, miR-470 and miR-134 as significantly upregulated upon retinoic acid treatment of mouse embryonic stem (ES) cells.[27] These miRNAs were also predicted to target the coding sequences of Nanog, Oct4 and Sox2, transcription factors known to be important in maintaining ES cells' pluripotency and in determining the initiation of differentiation.[28–31] Transfection of mouse ES cells with miR-296, miR-470 and miR-134 specifically downregulated levels of expression of Nanog, Oct4 and Sox2, and determined changes in cell morphology typical of differentiating ES cells; these results suggest that enforcing cross-species conservation and seed constraints might underestimate the number of possible miRNA targets.

Recent reports highlight the importance of truncations in the 3' UTR of genes, which can affect miRNA binding, as a novel cause of tumorigenesis.[32] For instance, *let-7* regulates the expression of the oncogene *HMGA2*, which encodes a chromatin-associated protein and contains seven *let-7* recognition sites in its 3' UTR. Chromosomal translocations in human cancers remove the *HMGA2* 3' UTR, thus probably contributing to tumorigenesis by disrupting *let-7*-mediated regulation.[32] Even in normal, healthy cells, about half of mammalian genes use alternative splicing and polyadenylation to generate multiple mRNA isoforms differing in the 3′ UTRs. Sandberg *et al.* performed a global analysis of alternative 3′ UTR isoforms during murine CD4+ T lymphocytes activation, and found that proliferating cells preferentially expressed mRNA isoforms with shortened 3′ UTR and fewer miRNA target sites.[33] It is tempting to speculate that UTR-based mRNA regulation plays distinct roles in the regulatory networks of non-proliferating or slowly-proliferating cells as compared to actively-proliferating cells, and that in some cases a shift towards expression of shorter, containing fewer miRNA target sites, 3′ UTRs may be required to evade regulation that would otherwise restrict cell cycle progression.[33]

5.4 Interplay Between miRNAs and Transcription Factors in Hematopoiesis

Transcription factors and miRNAs share a common regulatory logic, in which by binding to discrete *cis*-regulatory elements that are hard-wired on the DNA or mRNA sequence respectively, they can control expression of hundreds of target genes.[34] Recent advances in the field have begun to reveal how transcription factors and miRNAs can also regulate each other in molecular regulatory networks at the base of many biological processes.

In one study of genome-wide chromatin immunoprecipitation analysis, Zheng *et al.* have showed that the expression of FOXP3, a transcription factor that is required for the development and function of regulatory T (Treg) cells, may directly control the expression of *miR-155*, implicating miR-155 in Treg formation or function.[35] In another example of cross-talk between miRNAs and transcription factors, Taganov *et al.* performed promoter analysis of the human *miR-146a* locus and found that in monocytes the transcriptional upregulation of *miR-146a* induced by Toll-like receptor 4 is NF-κB dependent.[36] Since miR-146a was shown in this system to target TRAF6 and IRAK1, its upregulation determines a reduction in NF-κB activation in a negative feedback loop.[36] Conversely, the proto-oncogene *c-Myc* might be involved in the transcriptional repression of the *miR-146a* locus during tumorigenesis.[37] Finally, *in vitro* assays showed that the transcription factor PLZF interacts with and inhibits the *miR-146a* promoter, and that miR-146a targets CXCR4 mRNA, impeding its translation.[38] This cascade pathway controls megakaryopoiesis through PLZF suppression of *miR-146a* transcription and subsequent activation of CXCR4 translation.[38]

Expression of *miR-223* is regulated by a combination of factors. Initially, a circuit consisting of miR-223 as well as the transcription factors C/EBPα and NFI-A was described.[39] In this scenario, C/EBPα would activate transcription of *miR-223*, whereas NFI-A would repress it. By targeting NFI-A, miR-223 would turn off its repressor once it was expressed. This study suggested that sustained miR-223 expression is required for granulopoiesis. Further work has demonstrated that during granulocytic maturation the picture might be more complicated than this, as *miR-223* expression is also driven by myeloid transcription factors such as PU.1 and C/EBP, similar to many protein-coding genes involved in granulopoiesis.[40] Also, mice in which the *miR-223* gene was ablated showed an expanded granulocytic compartment.[41]

GATA-1 is a hematopoietic transcription factor essential for the formation of platelets, eosinophils, mast cells and erythrocytes.[42] Mice lacking GATA-1 die of severe anemia during embryonic development.[43] A recent study has shown that GATA-1 regulates expression of *miR-144* and *miR-451*. Moreover, loss of *miR-451* in zebrafish embryos caused severe anemia in a way that resembled (although less severely), the loss of *gata-1*, indicating that *miR-451* may be a major downstream effector of GATA-1 in erythroid cells.[44]

Another well-studied miRNA locus is the one encoding for the *miR-17-92* cluster. This is a cluster of seven miRNAs that is transcribed as a polycistronic unit and is highly homologous to the *miR-106a-92* and *miR-106-b* clusters located on different chromosomes. Monocytopoiesis is dependent on downregulation of the *miR-17-92* cluster, which allows expression of the transcription factor Runx1.[45] Accordingly, overexpression of miR-17, miR-20a and miR-106a suppressed Runx1 protein expression, leading to M-CSF receptor (M-CSFR) downregulation, enhanced blast proliferation and inhibition of monocytic differentiation and maturation. In turns, Runx1 was shown to be a transcriptional repressor of the *miR-17-92* and *miR-106a-92* clusters, demonstrating a mutually inhibitory miRNA-transcription factor circuit that regulates monocytic lineage differentiation.[45]

5.5 MiRNA Biogenesis' Role in Hematopoietic Development and Inflammation

The overall importance of miRNAs during hematopoiesis has been investigated by specific disruption of steps in miRNA biogenesis. During miRNA biogenesis, Dicer cleaves the pre-miRNAs into dsRNAs of ~22 nt. One strand of these dsRNA intermediates associates as mature miRNA in the RISC complex with a member of the Argonaute family of proteins. Argonaute proteins can be subdivided into Ago-like and Piwi-like subfamilies. While the Piwi-like proteins are specifically expressed during spermatogenesis and interact with piRNAs, the Ago proteins are broadly expressed in somatic cells and associate with miRNAs.[46,47] In humans, the *Ago* gene family consists of four members (*Ago1–Ago4*), however, only the Ago2 protein displays endonucleolytic or 'Slicer' activity.[48,49] Germline mutation of the *Ago2* gene resulted in embryonic lethality, while *Ago2* deficiency limited to the hematopoietic system resulted in impaired B cell and erythroid differentiation that triggered the population expansion of immature erythroblasts.[50] Retroviral rescue experiments surprisingly revealed that the Slicer endonucleolytic activity of Ago2 is dispensable for hematopoietic development as well as for miRNA biogenesis. Ago2 thus functions as a critical regulator of erythropoiesis and B cell development in a way that does not strictly require its catalytic activity. Interestingly, loss of *Ago2* also resulted in a global reduction of mature miRNAs in erythroblasts, fibroblasts and hepatocytes, indicating a Slicer-independent role for Ago2 protein in miRNA biogenesis.[50]

The RNAse III enzymes Drosha and Dicer mediate the stepwise maturation of miRNAs, and a general role for the miRNA pathway in regulating the differentiation and function of cells of the immune system has been demonstrated in mice lacking these enzymes.[51–57] MiRNA-dependent regulation is indeed critical for preventing spontaneous inflammation and autoimmunity, as genetic ablation of either Drosha or Dicer within the T cell compartment, or specifically within Foxp3+ regulatory T (Treg) cells, results in identical phenotypes.[51] Specifically, loss of Drosha throughout the entire T cell compartment resulted in spontaneous T cell activation, inflammatory disease and premature lethality. Deletion of either Drosha or Dicer solely in Foxp3+ Treg cells was even more devastating, and phenocopied Foxp3 or Treg deficiency.[51]

Conditional Dicer alleles avoid the lethality resulting from constitutive Dicer deficiency[58] and give insight into the role of Dicer-dependent RNAs in specific lineages. Dicer deletion early during T cell development induced by a *Cre* transgene driven by the *lck* promoter resulted in a sharp reduction of miRNAs by the double-positive (DP) stage and a 10-fold drop in the number of TCR-$\alpha\beta$ thymocytes.[53] Deletion of Dicer later in T cell development resulted in moderately reduced numbers of peripheral CD4+ T cells, which showed a higher tendency to differentiate into Th1-like, IFN-γ-expressing cells even when cultured under Th2-polarising conditions.[56] Thymic differentiation of Treg cells and FoxP3 expression were also compromised in the absence of Dicer and mature miRNAs, resulting in reduced Treg cell numbers and immune pathology.[52] Further studies have also shown that disruption of Dicer in Treg cells at the time of the first FoxP3 expression did not affect thymic Treg cell development, but the resulting cells adopted an effector-like phenotype with overlapping feature of Th1, Th2 and Treg cells.[57] These miRNA-deficient Treg cells presented altered cytokine profile and cell-surface expression of Treg cell markers, and failed to stably express FoxP3. Finally, mice with miRNA-deficient Treg cells rapidly developed spontaneous systemic autoimmune disease similar to *Foxp3*-null mice.[57] In another study that used a *Foxp3^{Cre}* knock-in allele to delete a conditional Dicer allele in Treg cells, the importance of miRNAs in Treg cells function was once again revealed especially under inflammatory conditions, where Tregs became activated but lost their suppressive capacity.[55] This led to the progression of fatal early onset lymphoproliferative autoimmune syndrome. Altogether, these data highlight the role of miRNAs for a stable Treg cell functional program.

Dicer deficiency in early B lymphocytes progenitors has been shown to block development at the pro- to pre-B transition.[54] To assess the role of Dicer on the V(D)J recombination program, B cell development was partially rescued by ablation of the pro-apoptotic molecule Bim, or by transgenic expression of the pro-survival protein Bcl-2. Dicer deficiency in these cells did not directly affect the V(D)J recombination program, but it affected antibody diversification.[54]

5.6 Control of Cell Survival and Cell Proliferation

Several miRNAs have been identified as important regulators of the cell cycle in different systems, and various studies have documented that the loss or gain of miRNA-mediated cell cycle control contributes to malignancy (reviewed in Ref. 59). MiRNA target proteins involved in cell cycle control include E2F transcription factors, cyclin-dependent kinase (Cdks), cyclins and Cdk inhibitors. Moreover, miRNAs themselves might be subject to cell cycle-dependent regulation, as it has been shown that translation regulation by some miRNAs can oscillate between repression and activation during the cell cycle.[60]

Using CLL-patient lymphocytes, Calin and colleagues identified the *miR-15a/miR-16-1* cluster as a candidate tumor suppressor gene, which was indeed downregulated in the majority of analyzed CLL samples.[61] Moreover *Bcl2*, an anti-apoptotic regulator whose protein product is over-expressed in CLL and other hematopoietic malignancies, was identified as a direct target of miR-15a and miR-16.[62]

During human erythropoiesis and erythroleukemic cell growth, miR-221–222 are downregulated, permitting production of the pro-survival surface receptor c-KIT and leading to an expansion of early erythroblasts.[63] Conversely, high levels of miR-221–222 in human glioblastomas and melanoma correlate with low levels of the cell cycle regulator p27^{Kip1} protein[64–66] and a higher proliferation rate. Mast cells are cells of the immune system that reside in most tissues and derive from hematopoietic precursors in the bone marrow.[67] Our lab recently identified miR-221–222 as important regulators of the cell cycle and cell proliferation in murine mast cells.[68] Overexpression of miR-221–222 in mast cells had a modest effect on the expression of the known target p27^{Kip1}; we showed that such partial effect on p27^{Kip1} in murine (but not human) cells was due to the expression of a splice variant of p27^{Kip1} that does not contain miR-221–222 binding sites in its 3′ UTR. More specifically, we found that the splice variant that possesses the miR-221–222 binding sites was primarily induced in mouse mast cells in the same conditions of stimulation that also induced miR-221–222 expression, indicating a possible role for miR-221–222 in regulating p27^{Kip1} upon induction, and therefore allowing p27^{Kip1} levels to return to basal levels after cell stimulation. These studies confirm that miRNA activity can be very dependent on the cellular environment, and that miRNA-mediated control can display specificity in terms of functional restriction to a particular cellular contest or differentiation pathway.

Indeed, because individual organisms can have hundreds of miRNAs and many thousands of miRNA targets, it is unlikely that the biological consequences of miRNA-mediated regulation will be the same in all cases. Some relevant parameters to consider are how much a given target is repressed by a miRNA, and how much the target repression matters in a given biological setting.[69] One given miRNA may not need to eliminate a target in order to have a substantial effect on the phenotype, if the target itself is highly dose-sensitive. Venturing into the arena of gene-dosage effect, the study from Xiao *et al.* showed that reducing by 25–35% the levels of c-MYB — by either removing one copy of the *c-Myb* gene or by overexpressing the *c-Myb*-regulator *miR-150* — resulted in developmental defects in the B cell lineage.[70]

5.7 Control of Cell Differentiation

Initially identified as a hematopoietic miRNA expressed most highly in the bone marrow,[6] miR-223 is expressed specifically in cells of the granulocytic lineage. Its expression changes during maturation, becoming higher as granulocytes mature.[41] Using a loss-of-function allele in mice, Johnnidis *et al.* showed that the myeloid-specific miR-223 negatively regulates progenitor proliferation and granulocytes differentiation and activation.[41] Mutant mice presented an expanded granulocytic compartment resulting from enhanced differentiation and proliferation of the granulocyte progenitor pool. In addition, granulocytes lacking miR-223 were hypersensitive to activating stimuli and displayed increased activity. As a consequence, *miR-223* mutant mice spontaneously developed inflammatory lung pathology and exhibited exaggerated tissue destruction after endotoxin challenge. The relevant target in this context seemed to be Mef2c, a transcription factor involved in promoting myeloid progenitor differentiation, as demonstrated by correction of the miR-223-null phenotype in mice lacking both *miR-223* and *Mef2c*.[41] Nevertheless, the correction of the phenotype was not complete, suggesting that additional miR-223 targets are involved. In these settings the physiological role of miR-223 would be therefore to fine-tune the generation and function of granulocytic cells, delimitating their production and dampening their activation.

MiR-150 is specifically expressed by mature lymphocytes and has a key role in B cell differentiation.[3,70] The expression of miR-150 increases during the maturation of B and T lymphocytes, but decreases rapidly when naïve T cells differentiate into Th1 or Th2 sub-types.[3] When ectopically expressed in hematopoietic precursors, miR-150 blocks B cell development at the transition from pro-B to pre-B, leading to severe defects in the production of mature B cells.[71] Similarly, knock-in mice with moderate ectopic but ubiquitous expression of miR-150 under the control of the *Rosa26* promoter showed a severe impairment in B cell development.[70] Impairment in T cell development was also present, but less pronounced. Mice in which the *miR-150* gene has been deleted show an increase in the number of splenic B-1 cells. B-1 cells reside mainly in the pleural and peritoneal cavities, and they are key participants in the early humoral response against invading pathogens. They often recognize self-antigens and common bacterial components, and antibodies produced by this B cell subset tend to be of broad specificity and low affinity.[70] The transcription factor c-Myb is a miR-150 predicted target; consistent with this observation, activated B cells deficient in miR-150 have more c-Myb, vice versa B cells overexpressing miR-150 have less c-Myb.[70] Importantly, subtle alterations in MYB levels — in the range of 25–35% — have profound effects on B cell development. In general, these results provide a rationale for why regulation of this target by a miRNA might have a particularly substantial impact on the phenotype, and support the concept that the physiological function of a given miRNA might lie in the control of the concentrations of just a few critical target proteins in particular cellular contexts. In addition, enforced expression of miR-150 in mouse hematopoietic progenitors resulted in considerable expansion of the megakaryocyte population at the expense of erythrocytes, indicating a role for miR-150 not only in lymphocytes function, but also in hematopoietic lineage commitment.[72]

The *miR-17-92* cluster is a polycistronic miRNA gene encoding six different miRNAs, miR-17, miR-18a, miR-19a, miR-20a, miR-19b and miR-92, some of which share identical 'seed' sequences. Two paralogues of the *miR-17-92* cluster (*miR-106b-25* and *miR-106a-363*) are apparently the result of a series of ancient genomic duplications and deletions.[73,74] Most of these miRNAs are widely expressed in various mouse tissues; in the lymphocyte lineage, the *miR-17-92* cluster is highly expressed in B and T precursor cells, and expression decreases after maturation.[74] Mice with targeted deletion of the *miR-17-92* cluster die within minutes of birth; fetal liver transplant experiments revealed a selective defect in B cell development, with a block at the transition from pro-B to pre-B and excessive apoptosis in the pro-B cell compartment, which correlates with increased levels of the pro-apoptotic protein Bim.[73] In contrast, mice with ectopic expression of the *miR-17-92* cluster in the lymphocyte compartment develop severe lymphoproliferative disease and autoimmunity and die prematurely.[74] Lymphocyte from these mice showed a lower threshold of activation after antigen recognition compared to control mice, and exhibited enhanced proliferation and reduced cell death, resulting in the accumulation of CD4+ T and B cells. Additionally, miR-17–92 has been shown to cooperate with *c-Myc* in a mouse model of B cell lymphoma.[75]

5.8 MiRNAs and Adaptive Immune Responses

One of the key features of a functioning immune system is its ability to distinguish antigens of foreign origin from those derived endogenously and to mount a productive immune response against the former. In addition to regulating hematopoietic cell lineage differentiation, miRNAs also have important roles in modulating adaptive immune responses in mice.

One of the first miRNAs that was shown to have a role in the development of vertebrate immune cells was miR-181a.[6] Specifically, *in vivo* transfer of bone marrow cells overexpressing miR-181a resulted in an increase in CD19+ B cells, a decrease in Thy1.2+ T cells and no change in myeloid cells.[6]

T lymphocytes sensitivity to antigens is regulated during thymocyte maturation to ensure proper development of immunity and tolerance. Li *et al.*[76] demonstrated that T cell receptor sensitivity and signaling strength is modulated post-transcriptionally by miR-181a. MiR-181a expression was higher in immature T cell populations, but lower in more differentiated

T cell populations such as Th1 and Th2 effector cells. Alteration of miR-181a levels of expression in these cells modified their sensitivity to antigens and impaired T cell selection. MiR-181a levels of expression therefore modulate T cell receptor signaling threshold, at least partially through downregulation of multiple phosphatases.[76]

The *bic* locus, a common retroviral integration site in ALV-induced (avian leucosis virus) B cell lymphomas, encodes the *bic* non-coding RNA, which is the primary transcript for miR-155. *Bic* and miR-155 are overexpressed in human B cell lymphomas, and mice expressing a *bic/miR-155* transgene develop B cell malignancies.[77] Consistent with the role of miR-155 in B cell responses, the expression of miR-155 was upregulated in this cell type upon activation.[78] Interestingly, mice deficient in the *miR-155* gene showed profound defects in their protective immune responses.[78,79] In these mice, the function of *miR-155*-deficient T and B lymphocytes, as well as dendritic cells was impaired, resulting in immunodeficiency. Providing mechanistic insights, the transcription factor PU.1 has been identified as a functionally relevant target of miR-155, negatively regulating switching to IgG1.[80] In addition, miR-155 targets the cytidine deaminase AID, a critical enzyme that mediates antibody class-switch recombination and somatic hypermutation.[81,82] MiR-155 was shown to have a role also in hematopoietic cell development, as enforced expression reduced both myeloid and erythroid colony formation of normal human CD34+ hematopoietic precursors,[83] although in another study, sustained expression of miR-155 in mouse bone marrow cells caused granulocyte/monocyte expansion, with pathological features characteristic of myeloid neoplasia.[84] While the exact mechanism by which miR-155 regulates hematopoiesis in these different models remains to be fully understood, these data highlight the importance of proper miR-155 expression in hematopoietic development and inflammation.

5.9 MiRNAs and Innate Immune Responses

The innate immune response, mediated by immune cells like granulocytes and macrophages, is the first line of defense against infections. This response is commonly mediated by activation of members of the Toll/ IL-1 receptor (TIR) superfamily. Signaling through the TIR cytoplasmic domain involves an association with the adaptor protein MyD88, which recruits IL-1 receptor-associated kinase 1 (IRAK1) and TNF receptor-associated factor 6 (TRAF6) following ligand binding.[85] Dissociation of IRAK1 from MyD88 following phosphorylation causes the activation of TRAF6, with subsequent initiation of a cascade that leads to the activation of transcription factors like NF-κB and AP-1 that in turn orchestrate the expression of immune response genes. To prevent an inappropriate inflammatory response following activation of the TIR receptors, a variety of feedback pathways have evolved to regulate this process. Perhaps not surprisingly, miRNAs are also emerging as regulators of innate immune responses.

MiRNA profiling demonstrated distinct patterns of miR-146 expression in various immune cell types, suggesting its involvement in the maintenance of lineage identity in immune cells subsets. Notably, miR-146 abundance was higher in murine Th1 cells but lower in Th2 cells relative to that in naïve T helper cells.[3] Most examined vertebrates have two copies of the gene encoding miR-146, *miR-146a* and *miR-146b*, located on separate chromosomes. These miRNAs share nevertheless the same 'seed' region and are therefore part of the same family. MiR-146a and miR-146b expression increases following activation of the innate immune response in monocytes/macrophages.[36] Importantly, miR-146a was found to be an NF-κB-dependent gene, and to regulate expression of TRAF6 and IRAK1.[36] Overexpression of miR-146 in the human monocytic cell line THP-1 could therefore fine-tune negative feedback regulation of inflammation through downregulation of IRAK1 and TRAF6. However, in a different system (lung alveolar epithelial cells) changes in miR-146a expression negatively regulated IL-1β-induced inflammatory response with a mechanism that did not seem to include IRAK1 and TRAF6 miRNA-mediated downregulation.[86] Even though the details of miR-146 mechanism of action still need to be further investigated, it is apparent that this miRNA is an important regulator of inflammation following activation of innate immune response.

Several reports have now indicated that some viruses encode their own miRNAs and that cellular miRNAs can positively or negatively influence viral replication.[87,88] Recently, it was found that one HCMV miRNA, miR-UL112, inhibits the translation of the cellular protein MHC class I-related chain B (MICB)[89] therefore protecting HCMV-infected cells against lysis by NK cells. This represented a novel miRNA-based immunoevasion mechanism exploited by HCMV, in which a viral miRNA directly down-regulated a host immune defense gene. Viral miRNAs can also act as orthologues of cellular miRNAs.[90] For example, miR-K12-11 encoded by Kaposi's-sarcoma-associated herpes virus (KSHV) shows significant homology to cellular miR-155 and regulates an analogous set of mRNAs. Even though it is still unclear which of the many genes regulated by miR-155 and miR-K12-11 provides a replicative advantage to KSHV, miR-K12-11 may contribute to the increased incidence of B-cell tumors seen in KSHV-infected patients.[77,91]

Epstein–Barr virus (EBV) was the first human virus found to encode miRNAs.[92] The EBV miRNAs are organized in two clusters within the EBV genome, one in the intronic region of the BART gene and the other in the untranslated regions of the BHRF1 gene. EBV miRNAs may be important not only for regulation of viral gene expression, but they may also target cellular transcripts to facilitate viral persistence and oncogenesis. Indeed, predicted cellular target of EBV miRNAs include regulators of cell growth and apoptosis, chemokines and cytokines, transcription factors and sig-naling molecules.[92] One of the cellular targets of the EBV miRNA miR-BART5 was recently shown to be the p53 upregulated modulator of apoptosis (PUMA), which can induce p53-independent apoptosis in response to a wide variety of stimuli.[93] These findings indicated that EBV encodes a miRNA that facilitate establishment of latent infection by pro-moting host cell survival.

Finally, it has been shown that some cellular miRNAs contribute to HIV-1 latency in resting T cells.[94] The latency of HIV-1 in resting primary CD4+ T cells is the major barrier for the eradication of the virus in patients on suppressive, highly-active, anti-retroviral therapy. It now appears that host cell miRNAs expressed in resting T cells potently inhibit HIV-1 production, suggesting that manipulation of cellular miRNAs could represent a novel approach for purging the HIV-1 reservoir. Overall, it is becoming clear that not only do many viruses encode their own regu-latory miRNAs to counteract immune responses, but host-encoded miRNAs have also been shown to interfere both positively and negatively with the viral life cycle.

5.10 Perspective

As a fundamental layer of post-transcriptional gene regulation, it is not surprising that miRNA genes have diverse and crucial roles in mammals. Not only do miRNAs have important roles in many cellular processes, but also their deregulated expression is observed in many diverse pathological conditions. Even though it is clear that miRNAs are essential components of the molecular circuitry that controls lymphocyte development, we are just scratching the surface of the full characterization of the role of miRNAs in the development and function of immune cells. Further studies on miRNA-mediated gene regulation during hematopoiesis and immune homeostasis will provide insights as to how aberrant miRNA expression may contribute to autoimmune diseases, leukemias and other immune-related pathological disorders.

Acknowledgments

I apologize to those whose work I was unable to cite. I would like to thank Vigo Heissmeyer for a critical reading of the manuscript. This work was supported by a Start-Up Grant from the IRB Foundation and by the Swiss National Science Foundation.

References

1. Kanellopoulou C. and S. Monticelli. (2008). A role for microRNAs in the development of the immune system and in the pathogenesis of cancer. *Semin Cancer Biol* **18**, 79–88.
2. Landgraf P., Rusu M., Sheridan R. *et al.* (2007). A mammalian microRNA expression atlas based on small RNA library sequencing. *Cell* **129**, 1401–1414.
3. Monticelli S., Ansel K.M., Xiao C. *et al.* (2005). MicroRNA profiling of the murine hematopoietic system. *Genome Biol* **6**, R71.
4. Neilson J.R., Zheng G.X., Burge C.B. *et al.* (2007). Dynamic regulation of miRNA expression in ordered stages of cellular development. *Genes Dev* **21**, 578–589.
5. Wu H., Neilson J.R., Kumar P. *et al.* (2007). MiRNA profiling of naive, effector and memory CD8 T cells. *PLoS ONE* **2**, e1020.

6. Chen C.Z., Li L., Lodish H.F. *et al.* (2004). MicroRNAs modulate hematopoietic lineage differentiation. *Science* **303**, 83–86.

7. Lee Y., Kim M., Han J. *et al.* (2004). MicroRNA genes are transcribed by RNA polymerase II. *EMBO J* **23**, 4051–4060.

8. Denli A.M., Tops B.B., Plasterk R.H. *et al.* (2004). Processing of primary microRNAs by the Microprocessor complex. *Nature* **432**, 231–235.

9. Gregory R.I., Yan K.P., Amuthan G. *et al.* (2004). The Microprocessor complex mediates the genesis of microRNAs. *Nature* **432**, 235–240.

10. Pillai R.S., Bhattacharyya S.N. and Filipowicz W. (2007). Repression of protein synthesis by miRNAs: How many mechanisms? *Trends Cell Biol* **17**, 118–126.

11. Obernosterer G., Leuschner P.J., Alenius M. *et al.* (2006). Post-transcriptional regulation of microRNA expression. *RNA* **12**, 1161–1167.

12. Thomson J.M., Newman M., Parker J.S. *et al.* (2006). Extensive post-transcriptional regulation of microRNAs and its implications for cancer. *Genes Dev* **20**, 2202–2207.

13. Heo I.C., Joo J., Cho M. *et al.* (2008). Lin28 mediates the terminal uridylation of let-7 precursor microRNA. *Mol Cell* **32**, 276–284.

14. Rybak A., Fuchs H., Smirnova L. *et al.* (2008). A feedback loop comprising lin-28 and let-7 controls pre-let-7 maturation during neural stem-cell commitment. *Nat Cell Biol* **10**, 987–993.

15. Viswanathan S.R., Daley G.Q. and Gregory R.I. (2008). Selective blockade of microRNA processing by Lin28. *Science* **320**, 97–100.

16. Davis B.N., Hilyard A.C., Lagna G. *et al.* (2008). SMAD proteins control DROSHA-mediated microRNA maturation. *Nature* **454**, 56–61.

17. Blow M.J., Grocock R.J., van Dongen S. *et al.* (2006). RNA editing of human microRNAs. *Genome Biol* **7**, R27.

18. Kawahara Y., Zinshteyn B., Sethupathy P. *et al.* (2007). Redirection of silencing targets by adenosine-to-inosine editing of miRNAs. *Science* **315**, 1137–1140.

19. Luciano D.J., Mirsky H., Vendetti N.J. *et al.* (2004). RNA editing of a miRNA precursor. *RNA* **10**, 1174–1177.

20. Yang W., Chendrimada T.P., Wang Q. *et al.* (2006). Modulation of microRNA processing and expression through RNA editing by ADAR deaminases. *Nat Struct Mol Biol* **13**, 13–21.

21. Kedde M., Strasser M.J., Boldajipour B. *et al.* (2007). RNA-binding protein Dnd1 inhibits microRNA access to target mRNA. *Cell* **131**, 1273–1286.

22. Lewis B.P., Burge C.B. and Bartel D.P. (2005). Conserved seed pairing, often flanked by adenosines, indicates that thousands of human genes are microRNA targets. *Cell* **120**, 15–20.
23. Lewis B.P., Shih I.H., Jones-Rhoades M.W. *et al.* (2003). Prediction of mammalian microRNA targets. *Cell* **115**, 787–798.
24. Miranda K.C., Huynh T., Tay Y. *et al.* (2006). A pattern-based method for the identification of microRNA binding sites and their corresponding heteroduplexes. *Cell* **126**, 1203–1217.
25. Baek D., Villen J., Shin C. *et al.* (2008). The impact of microRNAs on protein output. *Nature* **455**, 64–71.
26. Selbach M., Schwanhausser B., Thierfelder N. *et al.* (2008). Widespread changes in protein synthesis induced by microRNAs. *Nature* **455**, 58–63.
27. Tay Y., Zhang J., Thomson A.M. *et al.* (2008). MicroRNAs to Nanog, Oct4 and Sox2 coding regions modulate embryonic stem cell differentiation. *Nature* **455**, 1124–1128.
28. Loh Y.H., Wu Q., Chew J.L. *et al.* (2006). The Oct4 and Nanog transcription network regulates pluripotency in mouse embryonic stem cells. *Nat Genet* **38**, 431–440.
29. Mitsui K., Tokuzawa Y., Itoh H. *et al.* (2003). The homeoprotein Nanog is required for maintenance of pluripotency in mouse epiblast and ES cells. *Cell* **113**, 631–642.
30. Niwa H., Miyazaki J. and Smith A.G. (2000). Quantitative expression of Oct-3/4 defines differentiation, dedifferentiation or self-renewal of ES cells. *Nat Genet* **24**, 372–376.
31. Okita K., Nakagawa M., Hyenjong H. *et al.* (2008). Generation of mouse induced pluripotent stem cells without viral vectors. *Science* **322**, 949–953.
32. Mayr C., Hemann M.T. and Bartel D.P. (2007). Disrupting the pairing between let-7 and Hmga2 enhances oncogenic transformation. *Science* **315**, 1576–1579.
33. Sandberg R., Neilson J.R., Sarma A. *et al.* (2008). Proliferating cells express mRNAs with shortened 3' untranslated regions and fewer microRNA target sites. *Science* **320**, 1643–1647.
34. Hobert O. (2004). Common logic of transcription factor and microRNA action. *Trends Biochem Sci* **29**, 462–468.
35. Zheng Y., Josefowicz S.Z., Kas A. *et al.* (2007). Genome-wide analysis of Foxp3 target genes in developing and mature regulatory T cells. *Nature* **445**, 936–940.
36. Taganov K.D., Boldin M.P., Chang K.J. *et al.* (2006). NF-kappaB-dependent induction of microRNA miR-146, an inhibitor targeted to signaling proteins of innate immune responses. *Proc Natl Acad Sci USA* **103**, 12481–12486.

37. Chang T.C., Yu D., Lee Y.S. *et al.* (2008). Widespread microRNA repression by Myc contributes to tumorigenesis. *Nat Genet* **40**, 43–50.

38. Labbaye C., Spinello I., Quaranta M.T. *et al.* (2008). A three-step pathway comprising PLZF/miR-146a/CXCR4 controls megakaryopoiesis. *Nat Cell Biol* **10**, 788–801.

39. Fazi F., Rosa A., Fatica A. *et al.* (2005). A minicircuitry comprised of microRNA-223 and transcription factors NFI-A and C/EBPalpha regulates human granulopoiesis. *Cell* **123**, 819–831.

40. Fukao T., Fukuda Y., Kiga K. *et al.* (2007). An evolutionarily conserved mechanism for microRNA-223 expression revealed by microRNA gene profiling. *Cell* **129**, 617–631.

41. Johnnidis J.B., Harris M.H., Wheeler R.T. *et al.* (2008). Regulation of progenitor cell proliferation and granulocyte function by microRNA-223. *Nature* **451**, 1125–1129.

42. Ferreira R., Ohneda K., Yamamoto M. *et al.* (2005). GATA1 function, a paradigm for transcription factors in hematopoiesis. *Mol Cell Biol* **25**, 1215–1227.

43. Fujiwara Y., Browne C.P., Cunniff K. *et al.* (1996). Arrested development of embryonic red cell precursors in mouse embryos lacking transcription factor GATA-1. *Proc Natl Acad Sci USA* **93**, 12355–12358.

44. Dore L.C., Amigo J.D., Dos Santos C.O. *et al.* (2008). A GATA-1-regulated microRNA locus essential for erythropoiesis. *Proc Natl Acad Sci USA* **105**, 3333–3338.

45. Fontana L., Pelosi E., Greco P. *et al.* (2007). MicroRNAs 17-5p-20a-106a control monocytopoiesis through AML1 targeting and M-CSF receptor upregulation. *Nat Cell Biol* **9**, 775–787.

46. O'Donnell K.A. and Boeke J.D. (2007). Mighty Piwis defend the germline against genome intruders. *Cell* **129**, 37–44.

47. Peters L. and Meister G. (2007). Argonaute proteins: Mediators of RNA silencing. *Mol Cell* **26**, 611–623.

48. Liu J., Carmell M.A., Rivas F.V. *et al.* (2004). Argonaute2 is the catalytic engine of mammalian RNAi. *Science* **305**, 1437–1441.

49. Meister G., and Tuschl T. (2004). Mechanisms of gene silencing by double-stranded RNA. *Nature* **431**, 343–349.

50. O'Carroll D., Mecklenbrauker I., Das P.P *et al.* (2007). A Slicer-independent role for Argonaute 2 in hematopoiesis and the microRNA pathway. *Genes Dev* **21**, 1999–2004.

51. Chong M.M., Rasmussen J.P., Rudensky A.Y. *et al.* (2008). The RNAseIII enzyme Drosha is critical in T cells for preventing lethal inflammatory disease. *J Exp Med* **205**, 2005–2017.

52. Cobb B.S., Hertweck A., Smith J. *et al.* (2006). A role for Dicer in immune regulation. *J Exp Med* 203, 2519–2527.

53. Cobb B.S., Nesterova T.B., Thompson E. *et al.* (2005). T cell lineage choice and differentiation in the absence of the RNase III enzyme Dicer. *J Exp Med* 201, 1367–1373.

54. Koralov S.B., Muljo S.A., Galler G.R. *et al.* (2008). Dicer ablation affects antibody diversity and cell survival in the B lymphocyte lineage. *Cell* 132, 860–874.

55. Liston A., Lu L.F., O'Carroll D. *et al.* (2008). Dicer-dependent microRNA pathway safeguards regulatory T cell function. *J Exp Med* 205, 1993–2004.

56. Muljo S.A., Ansel K.M., Kanellopoulou C. *et al.* (2005). Aberrant T cell differentiation in the absence of Dicer. *J Exp Med* 202, 261–269.

57. Zhou X., Jeker L.T., Fife B.T. *et al.* (2008). Selective miRNA disruption in T reg cells leads to uncontrolled autoimmunity. *J Exp Med* 205, 1983–1991.

58. Bernstein E., Kim S.Y., Carmell M.A. *et al.* (2003). Dicer is essential for mouse development. *Nat Genet* 35, 215–217.

59. Chivukula R.R. and Mendell J.T. (2008). Circular reasoning: MicroRNAs and cell-cycle control. *Trends Biochem Sci* 33, 474–481.

60. Vasudevan S., Tong Y. and Steitz J.A. (2007). Switching from repression to activation: MicroRNAs can up-regulate translation. *Science* 318, 1931–1934.

61. Calin G.A., Dumitru C.D., Shimizu M. *et al.* (2002). Frequent deletions and down-regulation of micro-RNA genes miR15 and miR16 at 13q14 in chronic lymphocytic leukemia. *Proc Natl Acad Sci USA* 99, 15524–15529.

62. Cimmino A., Calin G.A., Fabbri M. *et al.* (2005). miR-15 and miR-16 induce apoptosis by targeting BCL2. *Proc Natl Acad Sci USA* 102, 13944–13949.

63. Felli N., Fontana L., Pelosi E. *et al.* (2005). MicroRNAs 221 and 222 inhibit normal erythropoiesis and erythroleukemic cell growth via kit receptor down-modulation. *Proc Natl Acad Sci USA* 102, 18081–18086.

64. Felicetti F., Errico M.C., Bottero L. *et al.* (2008). The promyelocytic leukemia zinc finger-microRNA-221/-222 pathway controls melanoma progression through multiple oncogenic mechanisms. *Cancer Res* 68, 2745–2754.

65. Galardi S., Mercatelli N., Giorda E. *et al.* (2007). miR-221 and miR-222 expression affects the proliferation potential of human prostate carcinoma cell lines by targeting p27Kip1. *J Biol Chem* 282, 23716–23724.

66. le Sage C., Nagel R., Egan D.A. *et al.* (2007). Regulation of the p27(Kip1) tumor suppressor by miR-221 and miR-222 promotes cancer cell proliferation. *EMBO J* 26, 3699–3708.

67. Galli S.J., Grimbaldeston M. and Tsai M. (2008). Immunomodulatory mast cells: Negative, as well as positive, regulators of immunity. *Nat Rev Immunol* 8, 478–486.

68. Mayoral R.J., Pipkin M.E., Pachkov M. *et al.* (2009). MicroRNA-221-222 regulate the cell cycle in mast cells. *J Immunol* **182**, 433–445.

69. Flynt A.S. and Lai E.C. (2008). Biological principles of microRNA-mediated regulation: Shared themes amid diversity. *Nat Rev Genet* **9**, 831–842.

70. Xiao C., Calado D.P., Galler G. *et al.* (2007). MiR-150 controls B cell differentiation by targeting the transcription factor c-Myb. *Cell* **131**, 146–159.

71. Zhou B., Wang S., Mayr C. *et al.* (2007). miR-150, a microRNA expressed in mature B and T cells, blocks early B cell development when expressed prematurely. *Proc Natl Acad Sci USA* **104**, 7080–7085.

72. Lu J., Guo S., Ebert B.L. *et al.* (2008). MicroRNA-mediated control of cell fate in megakaryocyte-erythrocyte progenitors. *Dev Cell* **14**, 843–853.

73. Ventura A., Young A.G., Winslow M.M. *et al.* (2008). Targeted deletion reveals essential and overlapping functions of the miR-17 through 92 family of miRNA clusters. *Cell* **132**, 875–886.

74. Xiao C., Srinivasan L., Calado D.P. *et al.* (2008). Lymphoproliferative disease and autoimmunity in mice with increased miR-17-92 expression in lymphocytes. *Nat Immunol* **9**, 405–414.

75. He L., Thomson J.M., Hemann M.T. *et al.* (2005). A microRNA polycistron as a potential human oncogene. *Nature* **435**, 828–833.

76. Li Q.J., Chau J., Ebert P.J. *et al.* (2007). miR-181a is an intrinsic modulator of T cell sensitivity and selection. *Cell* **129**, 147–161.

77. Costinean S., Zanesi N., Pekarsky Y. *et al.* (2006). Pre-B cell proliferation and lymphoblastic leukemia/high-grade lymphoma in E(mu)-miR155 transgenic mice. *Proc Natl Acad Sci USA* **103**, 7024–7029.

78. Thai T.H., Calado D.P., Casola S. *et al.* (2007). Regulation of the germinal center response by microRNA-155. *Science* **316**, 604–608.

79. Rodriguez A., Vigorito E., Clare S. *et al.* (2007). Requirement of bic/microRNA-155 for normal immune function. *Science* **316**, 608–611.

80. Vigorito E., Perks K.L., Abreu-Goodger C. *et al.* (2007). MicroRNA-155 regulates the generation of immunoglobulin class-switched plasma cells. *Immunity* **27**, 847–859.

81. Dorsett Y., McBride K.M., Jankovic M. *et al.* (2008). MicroRNA-155 suppresses activation-induced cytidine deaminase-mediated Myc-Igh translocation. *Immunity* **28**, 630–638.

82. Teng G., Hakimpour P., Landgraf P. *et al.* (2008). MicroRNA-155 is a negative regulator of activation-induced cytidine deaminase. *Immunity* **28**, 621–629.

83. Georgantas R.W., Hildreth III R., Morisot S. *et al.* (2007). CD34+ hematopoietic stem-progenitor cell microRNA expression and function: A circuit diagram of differentiation control. *Proc Natl Acad Sci USA* **104**, 2750–2755.

84. O'Connell R.M., Rao D.S., Chaudhuri A.A. *et al.* (2008). Sustained expression of microRNA-155 in hematopoietic stem cells causes a myeloproliferative disorder. *J Exp Med* **205**, 585–594.

85. Liew F.Y., Xu D., Brint E.K. *et al.* (2005). Negative regulation of toll-like receptor-mediated immune responses. *Nat Rev Immunol* **5**, 446–458.

86. Perry M.M., Moschos S.A., Williams A.E. *et al.* (2008). Rapid changes in microRNA-146a expression negatively regulate the IL-1beta-induced inflammatory response in human lung alveolar epithelial cells. *J Immunol* **180**, 5689–5698.

87. Cullen B.R. (2006). Viruses and microRNAs. *Nat Genet* **38 Suppl,** S25–S30.

88. Schutz S. and Sarnow P. (2006). Interaction of viruses with the mammalian RNA interference pathway. *Virology* **344**, 151–157.

89. Stern-Ginossar N., Elefant N., Zimmermann A. *et al.* (2007). Host immune system gene targeting by a viral miRNA. *Science* **317**, 376–381.

90. Gottwein E., Mukherjee N., Sachse C. *et al.* (2007). A viral microRNA functions as an orthologue of cellular miR-155. *Nature* **450**, 1096–1099.

91. Eis P.S., Tam W., Sun L. *et al.* (2005). Accumulation of miR-155 and BIC RNA in human B cell lymphomas. *Proc Natl Acad Sci USA* **102**, 3627–3632.

92. Pfeffer S., Zavolan M., Grasser F.A. *et al.* (2004). Identification of virus-encoded microRNAs. *Science* **304**, 734–736.

93. Choy E.Y., Siu K.L., Kok K.H. *et al.* (2008). An Epstein–Barr virus-encoded microRNA targets PUMA to promote host cell survival. *J Exp Med* **20**, 2551–2560.

94. Huang J., Wang F., Argyris E. *et al.* (2007). Cellular microRNAs contribute to HIV-1 latency in resting primary CD4+ T lymphocytes. *Nat Med* **13**, 1241–1247.

6

MicroRNAs Function as Tumor Suppressor Genes and Oncogenes

*Aurora Esquela-Kerscher**

MicroRNAs (miRNAs) control various processes related to cellular growth and differentiation and are commonly dysregulated in human cancers. A growing subclass of miRNAs is hypothesized to play an important role in cancer progression and function as tumor suppressor genes and oncogenes. We will discuss how miRNA misregulation in the diseased state can occur due to loss-of-function and activation mutations as well as due to abnormalities related to miRNA epigenetic silencing, transcriptional control, miRNA maturation, and miRNA-target recognition. We will also review accumulating experimental evidence indicating how specific miRNA-target interactions modulate tumor formation and metastasis. In some instances, a single miRNA controls opposing activities on tumorigenesis depending on the cellular context and environmental cues and highlights the complexity of miRNA-based regulation. Understanding the link between miRNA and cancer will provide insight into the utility of small RNAs as diagnostic and therapeutic tools for such a devastating disease.

*Department of Microbiology and Molecular Cell Biology, Eastern Virginia Medical School, 700 West Olney Road, Lewis Hall, Room 3047, Norfolk, VA 23507-1696, USA. E-mail: kerschae@evms.edu.

149

6.1 Introduction: MicroRNAs are Closely Linked with Cancer

Cancer is marked by uncontrolled proliferation and inappropriate survival of damaged cells in the body. Many processes used to direct the proper growth, differentiation and cell death of tissues in the developing embryo, are identical to the genetic pathways that are perturbed in the cancerous state. During the early stages of cancer progression, a cell acquires mutations in genes that regulate cellular growth and result in the inactivation of tumor suppressor genes and/or the induced expression of oncogenes. Much attention has been focused on a particular class of small ~22 nucleotide RNAs, microRNAs (miRNAs), due to the fact that the dysregulation of miRNA expression has been linked to human cancers and evidence indicates that these non-coding RNAs can also directly function as tumor suppressor and oncogenes.[1,2] MiRNAs bind to complementary sequences typically located in the 3′ untranslated region (3′ UTR) of their target protein-coding messenger RNAs (mRNAs) resulting in translational inhibition and/or mRNA degradation. There are estimated to be >700 miRNAs (and possibly as many as 1,000 miRNAs) in the human genome alone[3], comprising 1–3% of all expressed human genes and establishing miRNAs as one of the largest classes of gene regulators. Bioinformatic and *in vitro* studies imply that a single miRNA can regulate multiple targets (possibly exceeding 100), indicating that these small RNAs might control ~1/3 of all protein-expressing transcripts in the human genome. Therefore, the misregulation of miRNAs that direct cell cycle progression and differentiation could lead to the uncoordinated control of multiple cellular pathways and result in tumor formation.

This chapter will discuss the emerging link between miRNAs and cancer. Expression profiling studies have revealed that a large range of miRNA genes is dysregulated in human tumors. Although many of these small RNAs are likely mis-expressed due to secondary effects of the disease, a growing subclass of miRNAs is found to play a direct role in cancer progression. We will describe how the dysregulation of miRNAs related to cancer can result not only from loss-of-function and activation mutations of specific miRNA gene loci but also from abnormalities in miRNA epigenetic silencing, transcriptional control, miRNA processing and miRNA-target recognition. Select miRNA genes are considered *bona fide* tumor suppressor genes and oncogenes if they fulfill the following criteria: the miRNA is mis-expressed in primary human tumors compared to normal tissues; the miRNA directly influences cellular growth and differentiation processes in tissue culture as well as modulates tumor formation in animal models; the miRNA regulates the post-transcriptional expression of cancer-relevant targets. In addition, a number of miRNAs are associated with later stages of cancer progression, specifically processes related to metastasis. A specific example of each category will be summarized in this chapter. Understanding the relationship between miRNAs controlling cellular growth and differentiation pathways, and how their misregulation directs tumor formation and cancer progression, may lead to novel approaches for the diagnosis and treatment of the disease.

6.2 MiRNAs Control a Variety of Developmental Processes Related to Cancer Progression Pathways

Only a small fraction of the thousands of animal miRNAs identified to date have been assigned a biological function.[4] These miRNAs regulate essential developmental processes associated with cellular growth, differentiation and apoptosis, and therefore might act as tumor suppressor genes and oncogenes. For instance, the founding members of the miRNA family, *lin-4* and *let-7*, control the differentiation and proliferation of various cellular lineages in the nematode, *C. elegans*. *lin-4* and *let-7* were identified through genetic analysis to act as 'developmental switches' that control the timing of the larval transitions in *C. elegans*.[5–7] Loss-of-function mutations in the *lin-4* and *let-7* miRNAs result in reiterations of the first larval stage (L1) and the fourth larval stage (L4) fates, respectively. In wild-type animals, specialized skin cells, known as seam cells, divide at each larval stage with a stem cell-like pattern and terminally differentiate at the beginning of the adult stage. The seam cells fail to terminally differentiate in *lin-4* and *let-7* mutant animals, and instead reiterate the larval fate and divide again. Thus, defects in these miRNA genes result in cell cycle exit and differentiation defects, which are hallmarks of cancer.

There is accumulating evidence to support a conserved role for *lin-4* and *let-7* miRNAs in controlling processes related to cellular proliferation and differentiation. In Drosophila, the *lin-4* homologue, miR-125 and *let-7* are upregulated during the late larvae and prepupae stages,[8,9] and *let-7* has been shown to direct processes related to ecdysone hormone-induced metamorphosis.[10,11] Knockout studies in Drosophila indicate that *let-7* controls terminal cell-cycle exit programs in the wing imaginal discs during metamorphosis,[10] reminiscent of its role in programming terminal division of seam cells in the worm.[5] *let-7* is also found to control the timing of neuromuscular junction maturation within the fly abdominal muscles by downregulating the zinc-finger gene, *abrupt*, at the pupal stage.[10] Interestingly, the *lin-4* and *let-7* homologues are not expressed in human or mouse embryonic stem cells, embryonic carcinoma cells or during the earliest stages of mouse development but rather during later stages of development and in adult tissues, implying that these miRNAs possess a conserved function in mammals to induce cellular differentiation and cell cycle exit.[12–17] Expression studies for these small RNAs also imply a role in mammalian limb and brain development, and neuronal differentiation of EC and ES cells.[15–21] Indeed, the *lin-4* and *let-7* families regulate cell proliferation and differentiation in a variety of human cell lines as well as *in vivo* in mice and are mis-expressed in human cancers, clearly indicating a tumor suppressor function for these genes (discussed in section 6.4.1.2).[22–32]

Additional animal miRNAs have recently been identified to control cellular growth and differentiation pathways during development. The Drosophila miRNAs — *bantam*, miR-2, miR-6 and miR-14 — function to keep cell growth in check by controlling pathways related to cellular proliferation and apoptosis.[33–38] There are multiple examples of miRNAs that induce the specification of cellular populations to differentiate into certain lineages and organ structures. For example, *lsy-6* and miR-273 are involved in left–right asymmetric patterning of certain *C. elegans* sensory neurons;[39–41] miR-430 functions in zebrafish brain morphogenesis;[42] miR-375 functions in pancreatic islet-cell development and the regulation of insulin secretion;[43] miR-143 functions during mammalian adipocyte differentiation;[44] miR-196 functions in limb patterning;[45] miR-155 functions during T-cell and B-cell specification and the adaptive immune response;[46,47] the miR-1 family, miR-208, miR-133 and miR-195 all function during heart development;[48,49–52] and miR-181 is involved in T-cell differentiation and sensitivity to antigen,[53] B-cell lineage specification[54] and myoblast differentiation.[55] Since cancerous cells are hypothesized to revert back to a more embryonic fate as well as activate stem cell proliferative pathways, it is reasonable to propose that miRNA genes regulating developmental processes will be linked to those promoting cancer. Thus, further characterization of miRNAs that play essential roles during development and coordinate proper organ differentiation, embryonic patterning and body growth will facilitate our understanding of events associated with cancer progression and metastasis.

6.3 MicroRNA Biogenesis and Cancer

MiRNA biogenesis is a complex process that entails independent processing both in the nucleus and the cytoplasm before generating the biologically functional single-stranded ~22-nucleotide RNA species. MiRNAs are generally transcribed by Polymerase II resulting in a pri-miRNA precursor averaging 100–1,000s of nucleotides in length, which is often capped, spliced and polyadenylated, and can encode sequences for multiple miRNA genes. Within the nucleus, the Drosha microprocessor complex, composed of the RNase-III endonuclease Drosha together with the DiGeorge syndrome critical region gene 8 (DGCR8; or Pasha in Drosophila, PASH-1 in *C. elegans*), releases an ~70 nucleotide hairpin precursor termed a pre-miRNA. However, mirtrons, encoded in the introns of genes, can circumvent Drosha-mediated processing and generate pre-miRNA precursors directly from byproducts of intron splicing and disbranching events in the nucleus.[56] The pre-miRNA is exported out of the nucleus by the Exportin-5/Ran-GTP complex. Once in the cytoplasm, the pre-miRNA associates with a second RNase-III type enzyme Dicer and its co-factor, the mammalian TAR RNA binding protein (*TRBP*) (or the fly *loquacious* protein) resulting in the processing of a ~22 nucleotide miRNA–miRNA* duplex. Only one strand of the miRNA duplex is preferentially loaded into a large multi-protein miRNA ribonucleoprotein complex (miRNP, also referred to as the miRISC complex), which includes the Argonaute proteins, and modulates target gene expression by directing mRNA cleavage and/or translational inhibition.[57]

Factors responsible for proper miRNA transcription, processing and function have been implicated in cancer progression and tumor formation. For instance, oncogenes and tumor suppressor genes have been shown to bind to miRNA promoter elements and control their expression. The onco-genic transcription factor *c-Myc* associates directly with a conserved CATGTG sequence ~1.5 kb upstream of the miR-17-92 cluster and upregu-lates the transcription of the entire miRNA polycistron.[58] However, *c-Myc* can also repress the expression of at least 12 additional miRNA clusters including *let-7*, miR-15/16, miR-34a, miR-26a, miR-150, and miR-195/miR-497 when bound to their promoter elements,[59] reflecting the complex nature of miRNA regulation by this oncogene. Furthermore, the tumor-suppressor gene p53 can bind to the promoter regions of the miR-34 family (miR-34a and miR-34b/c) and activate their expression.[60–64] The overex-pression of miR-34 in primary and cancer cell lines leads to cell cycle arrest and apoptosis that is similar to the effects of p53 alone, and supports the notion that these miRNAs are themselves tumor suppressor genes.[60,61,63–66] Indeed, the miR-34 family is shown to repress the expression of targets related to cell cycle progression and cell death such as *N-MYC, CDK6, Cyclin D1, E2F3* and *BCL2*.[60,62,65,67,68] Therefore, cancerous cells exhibiting uncon-trolled growth and carrying mutations in protein-coding oncogenes and tumor suppressors might be transformed due to miRNA dysregulation.

Studies also suggest that loss of epigenetic control of certain miRNA loci located within CpG islands can contribute to cancer progression. For example, treatment of T24 bladder cancer cells with DNA methyltrans-ferase and histone deactylase inhibitors increases miR-127 expression dramatically and leads to a reciprocal downregulation of the *BCL6* proto-oncogene.[69] Hypermethylation of the miR-124 loci is frequently observed in colon, breast, and lung carcinomas, leukemias and lymphomas, and var-ious human cancer cell lines but not in normal tissues.[70] In lung cancer patients, miR-124a hypermethylation is linked to the upregulation of the oncogene, cyclin D kinase 6 (*CDK6*) and phosphorylation of the retinoblastoma (*Rb*) tumor suppressor. Furthermore, the *let-7a-3* locus associated with a CpG island is hypomethylated in a small subset of lung adenocarcinoma patients compared to extensive methylation of this gene in normal lung tissues.[71] The overexpression of *let-7a-3* in A549 lung can-cer cell lines can enhance their growth and reveals a potential oncogenic role for a member of the *let-7* family, which are generally classified as tumor suppressor miRNAs (Section 6.4.1.2). Finally, miR-34a is silenced due to CpG methylation in breast, lung, colon, bladder, pancreatic, and melanoma cell lines, as well as primary melanomas. The other miR-34 family members, miR-34b and miR-34c, are likewise reported to be CpG methylated in oral cell carcinoma and colorectal carcinomas.[72,73] It is estimated that 5–10% of mammalian miRNAs are epigenetically regulated[69,71,73–75], underscoring the importance of DNA methylation status and the control of gene expression at certain miRNA loci.

These findings illustrate the complexities of miRNA transcription in that miRNA promoters such as the miR-34 family are under both positive and negative control by multiple transcription factors, i.e., *c-Myc* and p53, as well as through epigenetic regulation. It is likely that a miRNA, which normally functions to modulate cellular growth and/or differentiation, can induce tumor formation when mis-expressed due to promoter dysregulation. Mis-expression of the tumor suppressor miR-34 has been linked to a variety of human cancers such as breast, lung, gastric, and prostate, and therefore factors controlling miRNA promoter transcription could potentially be used to treat these cancers.[23,75–77] The few regulatory factors identified to associate with miRNA promoters are themselves established tumor suppressors and oncogenes, and thus it will be important to characterize additional miRNA promoter elements more thoroughly in order to gain insight into how small non-coding RNAs contribute to human disease.

Proteins that control miRNA processing as well as components of the miRNA machinery necessary to direct target gene expression have also been implicated in cancer progression. The decreased expression of the processing enzymes Drosha and Dicer are observed in various cancers. For instance, Dicer expression is found to be downregulated in patients with non-small-cell lung carcinomas and strongly correlates with a shortened post-operative survival rate in these patients.[78] Reduced expression of both Drosha and Dicer mRNA levels is linked with more aggressive disease and a poor clinical outcome in patients with ovarian cancer.[79] The conditional deletion of *Dicer1* in a *K-RAS*-inducible mouse model results in enhanced cellular transformation and tumor formation in these animals.[80] Furthermore, truncating mutations in *TRBP*, the co-factor present within the Dicer complex and necessary for Dicer protein stabilization,[81] is observed in both hereditary and sporadic colon and gastric carcinomas with microsatellite instability.[82] These mutations block pre-miRNA processing and destabilize Dicer protein in colorectal cancerous cells; the reintroduction of *TRBP* in these cell lines result in tumor-suppressive properties such as decreased colony formation *in vitro* and lower tumorgenicity in nude mice.[82] It is likely that Dicer, *TRBP*, and Drosha play indirect roles in promoting cancer formation due to a global reduction of miRNA levels.

Additional factors that control miRNA processing have been implicated in human disorders. For instance, the RNA-binding protein *LIN-28* (which regulates miRNA processing primarily of the *let-7* miRNA family at both the Drosha and Dicer levels of biogenesis via binding to the precursor stem loop in order to block miRNA maturation)[15,21,83–86] has also been linked to cancer. This processing factor plays crucial roles in stem cell maintenance, primordial germ-cell development, and de-differentiation, and therefore it is not surprising that *LIN-28* mis-expression is also associated with human germ-cell tumors.[87,88] *LIN-28* levels are also noted to be elevated in human hepatocellular carcinomas and associated with a poor clinical outcome.[84] A more direct role for *LIN-28* in cancer progression was shown *in vitro*. Overexpression of *LIN-28* in human breast cancer MCF-7 cells can stimulate cancer cell proliferation as well as transform NIH/3T3 cell lines.[84,89] Since LIN-28 upregulation and *let-7* miRNA downregulation are often found in cancerous tissues, both of these factors might be useful diagnostic and therapeutic targets to treat human malignancies.

Argonaute proteins, which are crucial functional components of the miRNP complex, also have close associations with human cancer. There are four Argonaute proteins (Ago1–4/ eIF2C) found in the mammalian genome. Three of these proteins, *Ago3*, *Ago1*, and *Ago4* form a cluster on chromosome 1 (1p34–35) and are frequently deleted in Wilms' tumors of the kidney and linked with neuroectodermal cancers.[90,91] *Ago1* expression is increased in renal tumors lacking the Wilms' tumor suppressor gene, *WT1*,[90] as well as in the kidney during embryogenesis, suggesting a developmental role for *Ago1* in this tissue. Furthermore, *Ago2* is elevated in Estrogen Receptorα-negative (ERα-negative) breast cancer cell lines and primary breast tumors.[92] The overexpression of *Ago2* in ERα-positive MCF7 cells leads to a transformed phenotype that includes increased cellular proliferation and motility as well as decreased cell–cell adhesion.[92]

6.4 MiRNA Dysregulation and Disease

There is a strong correlation between miRNA mis-expression and cancer.[1] A large number of human miRNAs are located in 'fragile sites,' unstable areas within human chromosomes susceptible to deletions, amplifications, viral insertions and chromosome fusions, and are commonly associated with cancer.[23] For instance, *lin-4* and *let-7* miRNA homologues map to fragile sites often deleted in lung, breast, ovary, cervical, and urothelial cancers[23]. MiRNA expression is also found to be an accurate predictor of disease prognosis.[2] For example, patients with lower levels of *let-7* and higher levels of miR-155 expression in their lung carcinomas possess a worse post-operative survival rate than patients exhibiting reciprocal expression patterns.[28,93] Furthermore, miRNA expression profiling is shown to more accurately classify human cancers based on tissue origin than the expression profiling of protein-coding genes,[94] reflecting the potential use of miRNAs in clinical diagnosis. Profiling experiments of tumors from various human cancers has also revealed that there is widespread miRNA dysregulation in the diseased tissues compared to normal specimens, and the majority of these mis-expressed miRNA genes are downregulated.[24,94–99] This might reflect the dedifferentiation program of transformed cells and not necessarily a direct role for all of these small RNAs in cancer progression.[100] A subset of miRNAs are not only found to be poorly expressed or amplified in human cancers, but also control cancer progression pathways *in vitro* and possess the ability to directly modulate tumor growth *in vivo*. These small RNAs are considered tumor suppressor genes and oncogenes. Interestingly, certain miRNA genes sometimes direct converse effects depending on the cellular context and the environmental cues and therefore can play a dual role as both a tumor suppressor gene and an oncogene. This highlights the complexity of miRNA-based regulation and could reflect the vast array of target genes under miRNA control. Taken together, miRNAs have emerged as a major class of small RNAs that direct cancer progression pathways, and which have immense diagnostic and therapeutic potential. Although the mis-expression of many miRNA genes correlate with cancer progression, this section will specifically focus on miRNAs that possess substantial experimental evidence to support a biological role during tumorigenesis.

6.4.1 *MicroRNAs as tumor suppressors*

6.4.1.1 *MiR-15a and miR-16*

The human miRNAs, miR-15a and miR-16, clustered within the 13q14 locus, are located in the intron of a non-protein coding RNA gene of unknown function, called *LEU2*. Deletions in this 13q14 locus are found to occur in more than 65% of CLL cases, as well as in 50% of mantle cell lymphomas, 16–40% of multiple myelomas and in 60% of prostate cancers. MiR-15a and miR-16 are deleted or downregulated in > 50% of patients with a common form of adult leukemia, B cell chronic lymphocytic leukemia (CLL),[101] suggesting that these miRNAs function as tumor suppressor genes. In support of this notion, a point mutation in the *miR-16-1* precursor resulting in decreased miR-16 expression has been reported in two CLL patients.[102] The antiapoptotic gene, *BCL2*, which is often overexpressed in many types of human cancers including leukemias and lymphomas, is negatively regulated by miR-15a and miR-16.[103] Thus, loss or low levels of *miR-15* and *miR-16* are found to inversely correlate with increased *BCL2* and likely promote leukemo- and lymphomagenesis in hematopoietic cells.

Confirmation that miR-15a and miR-16 function as *bone fide* tumor suppressors was first revealed when miR-15a and miR-16 expression was restored in an acute megakaryocytic leukemia cell line, MEG-01, and found to induce apoptosis.[104] Furthermore, MEG-01 cells transfected with miR-15a and miR-16 and subsequently implanted into the flanks of immunocompromised mice and compared to control animals, results in dramatically inhibited tumor formation.[104] Microarray studies have indicated that these miRNAs also possess roles outside of the apoptosis pathway since mRNA transcripts downregulated following miR-15 and miR-16 delivery in cancer cell lines include genes associated with cell cycle progression and proliferation in addition to those related to antiapoptotic signaling. Indeed, transfection of miR-16 in colon, lung, breast, and ovarian cell lines lead to a block in the Go/G1 to S phase transition during the cell cycle by downregulating targets such as *CDK6*, *CDC27*, *CARD10* and *C10orf46*.[104,105] In a prostate cancer mouse model, miR-15a and miR-16 are found to control the proliferation, invasion and survival of prostate tumor cells by regulating the targets *BCL2*, Cyclin D1 (*CCND1*) and *WNT3A*.[106] Furthermore, inactivation of miR-15a and miR-16 activity within the intact wild-type mouse prostate results in marked prostatic hyperplasia. Inhibition of these miRNAs in non-tumorigenic prostate cell lines leads to tumor formation and enhanced tumor cell invasion when injected into immunodeficient mice.[106] Conversely, when miR-15a and miR-16 are overexpressed in human prostate cancer xenografts, the tumors are dramatically reduced.[106] These studies reflect the therapeutic potential of miR-15a and miR-16 in treating leukemia, prostate cancer and potentially, other types of cancers.

6.4.1.2 The let-7 family

The *let-7* miRNA family, consisting of 12 human homologues, is down-regulated in a large range of cancerous tissues and predicted to function as tumor suppressor genes. Certain members of the *let-7* family map to fragile sites on human chromosomes that are preferentially deleted and these sites are associated with lung, breast, ovary, urothelial, and cervical cancers.[23] For instance, *let-7g* is noted to map to 3p21, a well-known region deleted early in the progression of lung cancer. Early studies indicate that *let-7* could be a useful diagnostic tool for non-small cell lung carcinomas since patients who express lower levels of *let-7* in their lung tumors have a poorer prognosis and shortened postoperative survival compared to patients with higher *let-7* expression.[28,93] *In vitro* tissue culture experiments with human lung and liver cancer cell lines show that transient delivery of *let-7* reduces cellular proliferation and arrests cells in the G1-S transition in the cell cycle. Conversely, the inhibition of *let-7* activity leads to increased cellular division in these cell lines, indicating that *let-7* is a *bona fide* tumor suppressor gene.[26–28]

The *let-7* family negatively controls the expression of a wide range of targets that are associated with cellular proliferation and differentiation pathways. Members of the *let-7* family were initially shown to be important in controlling cell cycle exit and differentiation in *C. elegans* (Section 6.2) and found to genetically interact with the *let-60* gene, the worm orthologue of the *RAS* oncongene.[22,107] This genetic interaction is evolutionarily conserved and the *let-7* family controls *RAS* expression in human cells.[22] Since approximately 15–30% of all human tumors possess activating mutations in *RAS*, a membrane-associated GTPase signaling protein that regulates cellular growth and differentiation, it was hypothesized that the regulation of *RAS* by *let-7* would be disrupted in the diseased state. Indeed, lung cancer cell lines and primary lung tumors taken from patients with non-small cell carcinomas show a reciprocal expression pattern, in which *let-7* levels are downregulated and *RAS* is highly expressed in comparison to normal lung tissue.[22,28] Further analysis revealed that *let-7* negatively regulates additional oncogenes, *c-Myc* and *HMGA2*[30,108] as well as the cell cycle progression genes, Cyclin D2 (*CCND2*), *CDK6*, and *CDC25A*, implicating that this miRNA functions as a tumor suppressor gene by controlling multiple cancer progression pathways.[26] In fact, *let-7* is shown to silence *H-RAS* resulting in reduced self-renewal and to repress *HMGA2* leading to enhanced differentiation of breast cancer stem cells *in vitro*.[29] Overexpression of *let-7* in these highly-tumorigenic breast cancer stem cells significantly reduces tumor formation and lung and liver metastasis when implanted into immunocompromised mice.[29] Likewise, *let-7* is also shown to dramatically reduce lung tumor formation *in vivo* when virally delivered to mice that form lung adenocarcinomas due to the expression of an activating *K-RAS* mutation.[27,31] Taken together, these studies indicate that *let-7* is a promising therapeutic agent to treat cancers by repressing the expression of multiple targets that regulate cellular proliferation and differentiation.

6.4.1.3 *MiR-26a*

MiR-26a is a potent tumor suppressor gene that is under complex transcriptional control. The *c-Myc* oncogene is shown to bind to the miR-26a promoter and repress its expression. Furthermore, in a c-*Myc*-driven B lymphoma mouse model, enforced expression of miR-26a leads to dramatic reduction in tumorigenesis.[59] MiR-26a is also regulated at the promoter level by the tumor suppressor p53 and miR-26 expression is activated by p53 upon DNA damage.[61,109] Downregulation of miR-26a has also been linked to anaplastic thyroid carcinomas[110] and hepatocellular carcinomas;[111] deletion of 3p21.3 containing the *miR-26a-1* locus is associated with small cell lung carcinomas, renal cell carcinomas, and breast carcinomas.[112] *In vitro* studies in liver cancer cells (HepG2) reveal that miR-26 regulates cellular proliferation at the level of cell cycle progression.[111] Overexpression of miR-26a in HepG2 cells led to cell cycle arrest at G1 due to miRNA-mediated repression of cyclin D2 (*CCND2*) and cyclin E2 (*CCNE2*). *In vivo* analysis also supports a tumor suppressor role for miR-26a in the liver.[111] Mice engineered to express an inducible form of the *Myc* oncogene specifically in the liver for ten weeks exhibit fulminant liver disease, in which the liver is almost completely replaced with tumors. However, the majority of animals delivered an adeno-associated virus (AAV) expressing miR-26a seven weeks following *Myc* induction show small liver tumors or complete absence of tumors compared to the controls. Tumor reduction of the miR-26 treated mice occurs due to an inhibition of cell proliferation as well as by increased apoptosis, although apoptosis-associated targets for miR-26 have yet to be identified. Interestingly, only tumor cells in this model are susceptible to miR-26-mediated apoptosis and normal livers overexpressing miR-26 appear to be unaffected. These studies hold great promise for systemically treating liver cancer with miRNAs such as miR-26 that are highly expressed in normal tissues but underexpressed in tumors.

Although these studies firmly establish a tumor suppressor role for miR-26a in the liver, there is evidence that this small RNA can function as an oncogene in the glia and is overexpressed in a subset of high-grade human gliomas, resulting primarily from amplifications of the *miR-26a-2* locus.[113] MiR-26 represses the expression of the tumor suppressor gene *PTEN* (phosphatase and tensin homolog) *in vitro* and miR-26a-mediated repression of *PTEN* is found to enhance tumor formation in a mouse glioma model.[113] Therefore, cellular context and tissue-specific expression of miRNA targets likely dictate if miR-26a functions as a tumor suppressor gene (liver) or an oncogene (glia) — a common theme that will be encountered for additional miRNAs, such as the miR-29 family and the miR-17-92 cluster described below in Section 6.4.1.4.

6.4.1.4 *The miR-29 family*

The miR-29 family, consisting of two clusters, *miR-29b-1/miR-29a* (7q32) and *miR-29b-2/miR-29c* (1q23), are downregulated in chronic lymphocytic leukemia (CLL), lung, colon, and breast cancer, acute myeloid leukemia (AML) and cholangiocarcinomas.[24,93,102,114,115] This miRNA family can function as tumor suppressors both *in vitro* and *in vivo* in a variety of cell types. For example, when malignant cholangiocarcinoma cells, HeLa cervical cancer cells, or MCF-7 breast cancer cells are transfected with miR-29b, these cells lines show enhanced apoptosis and in the case of cholangiocarcinoma cells, a sensitized response to TRAIL cytotoxicity.[115,116] This response is mediated by the repression of the miR-29 target, *MCL-1*, an antiapoptotic *BCL2* family member[115] as well as by directly activating the p53 pathway by downregulating the target genes, p85alpha (the regulatory subunit of *PI3* kinase) and *CDC42* (a Rho family GTPase).[116] Likewise, ectopic expression of the miR-29 miRNAs in A549 lung cancer cells inhibits cellular growth and induces apoptosis *in vitro*. When A549 cells overexpressing miR-29 miRNAs are implanted into immunocompromised mice, these miRNAs dramatically inhibit lung tumor growth.[117] The antineoplastic effects of the miR-26 family can be potentially attributed to a multitude of targets related to cell survival and epigenetic status such as the downregulation of apoptosis related targets, Mcl-1, p85alpha and *CDC42*, as well as the T-cell leukemia 1 (TCL-1) oncogene[118] and the DNA methyltransferases (*DNMT*) 3A and -3B.[117] In addition, overexpression of miR-29 miRNAs in rhabdomyosarcoma cells also show a marked decrease in cell proliferation and increased differentiation when injected into mice possibly by targeting *Yin-Yang-1* (*Yy1*).[119]

However, the miR-29 family also appears to have a tumor- and metastasis-promoting role in breast cancer cells by suppressing the expression of tristetraprolin (*TTP*), a protein involved in the degradation of mRNAs containing AU-rich 3′ UTRs and functions as an important regulator of inflammation, epithelial-to-mesenchymal transitions (ETMs), and metastasis.[120] Murine mammary epithelial cells engineered to overexpress the *RAS* oncogene (EpRas) and also overexpress miR-29 are found to have increased levels of lung metastasis when injected into the tail vein of immunodeficient mice compared to cells lacking miR-29 expression.[120] The increased levels of lung metastasis are due to miR-26-mediated repression of *TTP* and knockdown of this protein in EpRas cells lead to similar frequencies of lung metastasis as observed for cells overexpressing miR-29. Furthermore, these EpRas mammary cells show disrupted epithelial polarity in culture when overexpressing miR-29 or depleted of the target *TTP*. As predicted, human breast cancer specimens that exhibit invasive tumors show greatly enhanced levels of miR-29 and decreased *TTP* compared to samples of benign, non-invasive breast hyperplasias.[120]

These results indicate that miR-29 directs opposing cellular functions and therefore, tissue context, cellular cues and disease state likely dictate if this miRNA suppresses or promotes cancer progression pathways.

6.4.2 Oncogenic microRNAs

6.4.2.1 MiR-155

The *BIC* locus works cooperatively with the *c-Myc* oncogene, a basic region-helix-loop-helix-leucine zipper transcription factor that controls cellular proliferation, growth and cell death, and together they induce B-cell lymphomas.[121] *BIC* is a non-coding RNA that harbors a highly conserved region of 138 nucleotides that encodes the hairpin region of miR-155.[121] This locus was initially identified as a common integration site for the avian leucosis virus in chicken lymphoma cells and is also often overexpressed in human B-cell lymphomas.[122,123] MiR-155 expression is upregulated in pediatric Burkitt lymphoma, Hodgkin lymphoma, and in primary mediastinal large B-cell and diffuse large B-cell lymphomas, reflecting a role for this gene in hematopoiesis.[124–126] Indeed, miR-155 is expressed in mature B and T lymphocytes, germinal center B cells and activated monocytes.[47]

Gene disruption studies in mice for miR-155 also support a role for this miRNA in the immune system.[46,47] Mice lacking miR-155 throughout development exhibit normal B cell proliferation but show defects related to germinal center response at the level of cytokine production, decreased dendritic cell function and a CD4+ T cell bias towards Th2 differentiation.[46,47] These animals also present extensive lung remodeling and enteric inflammation, indicating a role in overall immune system homeostasis.[46] However, miR-155 is also upregulated in solid tumors such as breast, lung, colon, and pancreatic cancers, which implies additional functions for this gene in other organ systems.[24,99,127,128]

The overexpression of miR-155 specifically in the B-cells of transgenic mice results initially in preleukemic pre-B cell polyclonal proliferation in the spleen and bone marrow and later in the development of lymphoblastic leukemia and high-grade lymphomas.[129] These results reveal a direct oncogenic role for miR-155 and imply that miR-155 is responsible for the initiation of these malignancies but progression of the disease requires additional genetic alterations to allow for a full transformation to occur in these transgenic animals. Little is known regarding the targets required by miR-155 in order to regulate immunological responses during development as well as to mediate cancer progression. The few targets identified include the transcription factor *MAF*,[47] the transcriptional repressor involved in hypoxia signaling *BACH1*,[130,131] and the activation-induced cytidine deaminase *(AID)*[132,133] that is required for immunoglobulin gene diversifications and chromosomal translocations.

6.4.2.2 *The miR-17-92 cluster*

A non-coding RNA, *c13orf25*, mapping to the 13q31 locus harbors the *miR-17-92* cluster, which includes six miRNAs — *miR-17, miR-18a, miR-19a, miR-20a, miR-19b-1* and *miR-92-1* — within an 800-basepair region. The genomic locus 13q31 is preferentially amplified in hematopoietic malignancies such as diffuse large B-cell lymphoma, follicular lymphoma, mantle cell lymphoma, and primary cutaneous B-cell lymphoma, as well as in alveolar rhabdomyosarcoma and liposarcoma.[134–136] The *miR-17-92* cluster is hypothesized to possess oncogenic characteristics since members of this cluster are upregulated in 65% of B-cell lymphoma samples tested, are elevated in solid cancers, e.g., breast, colon, lung, pancreas, prostate, and stomach, and this locus is a common retroviral insertion site in murine leukemias.[97,99,137–141] Knockout studies in mice reveal that miR-17-92 play important roles in lung, heart, and B cell development and lymphomagenesis.[142] Mice lacking *miR-17-92* function die shortly after birth and exhibit ventricular septum defects, lung hypoplasia, and decreased numbers of circulating lymphocytes due to a significant reduction of pre-B and mature B cells via apoptotic mechanisms. Furthermore, miR-17-92 functions redundantly with one of two highly conserved clusters, namely miR-106b-25 (but not with miR-106a-363). When the miR-17-92 and miR-106b-25 clusters are deleted in combination, the resulting mutant mice die at mid-gestation and exhibit severe apoptosis defects during heart and B cell development as well as in regions of the central nervous system and the liver. Based on these mouse disruption studies, it is not surprising that the misregulation of the miR-17-92 cluster is linked to a wide array of tissues.

Functional analysis of miR-17-92 reveals that overexpression of a truncated miR-17-19b-1 cluster (lacking miR-92-1) can accelerate the formation of malignant lymphomas in a *Myc* transgenic mouse model.[143] Lymphomas arising due to elevated levels of both the *Myc* oncogene and the truncated miR-17-19b-1 cluster are found to have increased cellular proliferation rates and decreased apoptosis compared to non-aggressive malignancies arising from increased *Myc* alone, which exhibit extensive cell death. Interestingly, individual miRNAs of the truncated miR-17-19b-1 cluster fail to accelerate disease onset, implying that certain miRNAs within the cluster are required cooperatively to induce an oncogenic effect in this model. In support of this notion, it is shown that mice overexpressing the miR-17-92 cluster specifically in B and T cells, and at levels comparable to those observed in human lymphoma cell lines, result in lymphoproliferative disease and autoimmunity defects.[144]

As mentioned previously, the *Myc* oncogene directly activates the transcription of the miR-17-92 cluster and *Myc* binds to DNA sequences located within the first intron of the c13orf25 gene.[58] In turn, the miR-17-92 cluster negatively regulates the expression of the transcription factor *E2F* family, *E2F1*, *E2F2* and *E2F3*, which controls the G1 to S phase transition during the cell cycle by controlling genes associated with DNA replication, cellular proliferation, and apoptosis.[58,145–147] *E2F1* and *E2F3* can also activate the expression of the miR-17-92 cluster and function in a negative feedback loop with these miRNAs.[137] Given that *E2F1*, *E2F3*, and *Myc* directly activate the transcription of the miR-17-92 cluster and the E2F family independently functions in a reciprocal positive feedback loop with *Myc*, an intricate signaling network between *Myc*, the miR-17-92 cluster, and the *E2F* family exists to fine-tune cellular growth and apoptosis. Furthermore, excessive levels of *E2F* proteins (most notably *E2F1*) induce apoptosis whereas moderate *E2F* levels drive cellular proliferation.[145] Thus, in a model supporting an oncogenic role for the miR-17-92 cluster, the negative regulation of *E2F1* by the miR-17-92 cluster blocks *E2F*-mediated apoptosis and promotes *Myc* activated cellular proliferation. The miR-17-92 cluster is also shown to negatively regulate addition targets related to cancer progression pathways, which further supports an oncogenic role for these small RNAs. For instance, miR-17 targets the cyclin-dependent kinase inhibitor *CDKN1A* (p21),[148] a negative regulator of cell cycle progression at G1, certain cluster members target the pro-apoptotic gene *BCL2L11/BIM*,[137,142,144,149] and the miR-17-92 cluster has been implicated in the downregulation of the antiangiogenic proteins thrombospondin-1 (*TSP1*) and connective tissue growth factor (*CTGT*), as well as the tumor suppressor phosphatase and tensin homologue (*PTEN*).[144,150]

There is also evidence that the miR-17-92 cluster can function to suppress tumor formation. In this scenario, the miR-17-92 cluster acts to dampen *E2F* activity and block the cell proliferative effects of *Myc* by breaking the positive feedback loop between *Myc* and *E2F*. Consistent with this model, loss-of-heterozygosity of the 13q31 loci has been reported for multiple tumor types such as hepatocellular carcinoma; deletions in the miR-17-92 cluster are reported in 16.5% of ovarian cancers, 21.9% of breast cancers, and 20% of melanomas; and miR-17 can inhibit the proliferation of breast cancer cells *in vitro*.[151–153] The dual role of the miR-17-92 cluster as tumor-suppressing facors and oncogenes is similar to that discussed for miR-26 and the miR-29 family and underscores the complexities of tumorogenesis.

6.4.2.3 *MiR-21*

MiR-21 is found to be upregulated in virtually every cancerous cell line and tumor type analyzed, e.g., breast, colon, lung, pancreas, stomach, and prostate, and is predicted to function as an oncogenic miRNA.[24,99,154] Given that miR-21 upregulation in diseased cells rarely correlates with amplification of the miR-21 genomic locus at 17q23.2, deregulation of this RNA must occur at the level of transcription or by post-transcriptional events such as miRNA processing. MiR-21 was initially found to be dramatically upregulated in highly malignant glioblastomas.[155,156] Inactivation of miR-21 in glioblastoma cells, as well as in breast and prostate cell lines, reveals that this miRNA controls cellular growth primarily by inhibiting apoptosis in a variety of tissues.[155,157,158] These effects are also observed *in vivo*. For instance, when breast cancer cells transfected with a miR-21 inhibitor are implanted into the mammary pads of mice, the resulting tumors are half the size of the untreated tumors.[158] Interestingly, miR-21 has differential biological effects depending on the cell lines analyzed *in vitro*. For example, when similar inhibition assays are performed for miR-21 in highly tumorigenic and metastatic breast (MDA-MB-231) and prostate (PC-3MM) cancer cell lines, miR-21 shows no effect on cellular growth, and rather causes decreased cellular invasiveness and motility *in vitro* and dramatic reduction of overall metastasis in mouse xenografts.[159] A number of targets have been identified that are downregulated by miR-21 and associated with processes related to tumor growth, apoptosis, and metastasis. Targets for miR-21 include the antiapoptotic protein *BCL2*, the programmed cell death protein 4 (*PDC4*), tropomyosin 1 (*TPM1*) and the tumor suppressor gene maspin (*SERPINB5*),[158–160] all of which act to suppress cellular growth, invasion, and metastasis; the metalloproteinase inhibitors reversion-inducing cysteine-rich protein with kazal (*RECK*) and tissue inhibitor of metalloproteinases (*TIMP3*),[161] as well as Sprouty 2 (*SPRY2*) that regulates cellular outgrowth, branching, and migration.[162] Indeed, miR-21 might prove useful as a prognostic marker for a wide range of cancers. Higher levels of miR-21 are detected in more aggressive and malignant tumors compared to early disease stages for gliomas, breast cancer, colon adenocarcinomas, pancreatic cancer, and gastric carcinomas, and this trend is often associated with poor patient prognosis.[163]

6.5 A Role for MicroRNAs in Metastasis

Metastasis is a complex process in which cancerous cells from the original tumor site attain characteristics that allow them to detach from the tumor, recruit and migrate to blood and lymph vessels, invade into the vasculature network, travel and survive in the blood to a distant site, exit the vasculature and ultimately establish new tumor growth. MiRNAs have recently been shown to control these processes related to advanced stages of cancer progression and tumor metastasis, examples of which are discussed in this section.

6.5.1 *MicroRNA activators of metastasis*

6.5.1.1 *MiR-10b*

MiR-10b is upregulated in metastatic breast cancer cell lines compared to primary breast cancer and normal breast tissues, implicating a role for this small RNA in later cancer events.[164] Indeed, miR-10b is shown to induce cell migration and invasion of breast cancer cells both *in vitro* and *in vivo*.[164] The overexpression of miR-10b in immortalized and non-metastatic breast cancer cell lines results in a marked increase in cell motility and invasive characteristics but has no effect on their proliferation rates. Conversely, inhibition of miR-10b activity leads to a 10-fold reduction of cellular invasion but not viability in metastatic cell lines. Furthermore, implantation of a non-metastatic breast cancer cell line overexpressing miR-10b into the mammary fat pads of immunocompromised mice results in tumors that often invade into the surrounding stroma, musculature, and vasculature as well as to distant sites such as the lung. Closer analysis of the resulting tumors overexpressing miR-10b reveal invasion fronts with elevated levels of cellular proliferation and angiogenesis. In contrast, control non-invasive cell lines form tumors following animal implantation that never exhibit invasive behavior. Two regulatory factors have been identified to explain the metastatic potential observed for miR-10b.[164] The transcription factor *TWIST1*, previously found to control the epithelial-to-mesenchymal transitions (EMTs) of cancer cells contributing to the invasive characteristics during metastasis, is shown to directly bind to the miR-10b promoter and induce miRNA expression. Silencing of miR-10b in breast cancer cell lines overexpressing *TWIST1* show a reduction in motility and invasive behavior, however, unlike *TWIST1*, miR-10b is unable to induce EMT. Homeobox D10 (*HOXD10*) is a *bona fide* target for miR-10b and functions as a potent inhibitor of cellular migration and invasion *in vitro* primarily by repressing the pro-metastatic RHOC protein and can block tumor progression in mice. Interestingly, the *miR-10b* gene is located in the *HOXD* cluster and implies a developmental role for this miRNA in embryonic patterning.

6.5.1.2 *MiR-373 and miR-520c*

In a screen of 450 miRNA genes to identify small RNAs that can stimulate cell migration in a non-metastatic, non-migratory breast cancer cell line, both miR-373 and miR-520c were shown to have positive effects on this metastatic process.[165] In fact, these two miRNAs can activate both migration and invasion of breast cancer cell lines *in vitro* but have no effect on cellular proliferation or cell cycle progression. Both miR-373 and miR-520c can direct potent metastatic effects *in vivo* when overexpressed in non-invasive MCF-7 breast cancer cells and injected into immunodeficient mice. Six to eight weeks following injection, breast cancer cells overexpressing either miRNA formed metastatic nodules in the skull, spine, and jaw, resembling pathology observed in human breast cancer bone metastasis as well as in the lung pleura that are never observed in the control cells. However, the ability to stimulate cell migration in cancer cells of prostate and colon origin is only found to extend to miR-373 but not miR-520c and indicates that the tissue-specific nature of miRNA function should be carefully considered.

The cell surface receptor for hyaluronan, *CD44*, is a direct target for both miR-373 and miR-520c[165] and functions as a metastatic suppressor in prostate and colon cancer. In fact, decreased expression of *CD44* in metastatic breast cancer is linked to poor clinical outcome.[166–169] As predicted, non-invasive breast cancer cell lines such as MCF-7 exhibit high levels of *CD44* and low expression of miR-373 and miR-520c. Furthermore, either the overexpression of miR-373 and miR-520c or the suppression of *CD44* expression via RNAi in non-invasive MCF-7 cell lines similarly results in the induction of skull and lung metastasis *in vivo*. These metastases are dramatically reduced in MCF-7 cells that both overexpress miR-373 and miR-520c as well as carry a truncated version of *CD44* lacking a 3′ UTR (and thus miR-373 and miR-520c binding sites), as would be expected if a *bone fide* miRNA-target interaction exists. An inverse relationship between miR-373 and *CD44* is also observed in primary tumor and lymph node metastases of breast cancer patients.[165] MiR-373 expression levels are significantly higher and *CD44* expression greatly reduced in the primary tumors of patients with lymph node metastases compared to lymph-node negative patients, indicating a potential use of these two genes as future biomarkers for breast cancer prognosis. Interestingly, miR-373 is implied to function as an oncogene in cooperation with *H-RAS* in testicular germ-cell tumors by blocking p53-mediated pathways and inhibiting the tumor suppressor gene, *LATS2*.[170] However, the transformation ability of miR-373 in this context appears to be independent of its metastatic role in breast cancer cell lines.[165]

6.5.2 *MiRNA suppressors of metastasis*

6.5.2.1 *MiR-335, miR-126 and miR-206*

MiR-335, miR-126, and miR-206 were the first miRNAs identified to specifically block metastatic pathways. These miRNAs are expressed at low levels in various human metastatic lung and bone breast cancer cell lines, and decrease both lung and bone metastasis formation in animal models when selectively overexpressed compared to untreated metastatic cells.[171] MiR-126 is found to inhibit cellular proliferation *in vitro* and the overall growth and proliferation of tumors *in vivo* without inducing apoptosis.[171] Conversely, miR-335 and miR-206 have no effect on proliferation or apoptosis of the metastatic tumors, and rather reduce migration and invasion of breast cancer cells *in vitro*.[171] Therefore, these miRNAs suppressed metastasis using distinct cellular mechanisms and presumably downregulate distinct mRNA targets. These miRNAs also correlate with breast cancer disease prognosis. Low expression levels of miR-335, miR-126 and miR-206 are observed in primary tumors of patients with shorter median time of metastatic relapse to lung, bone, or brain. Furthermore, reduced miR-335 and miR-126 expression in primary breast cancer tumors correlate with a poor prognosis of metastasis-free survival.

The targets controlled by these miRNAs are beginning to be identified and, as expected, they are associated with processes involved in metastasis. For example, miR-335 negatively regulates the *SRY*-box containing transcription factor (*SOX4*) (a gene associated with progenitor cell development and cell migration), and tenascin C (*TNC*) (a regulator of cell migration), as well as genes linked to extracellular matrix and cytoskeleton control, specifically the receptor-type tyrosine protein phosphatase *PTPRN2* and the c-*MER* tyrosine kinase (*MERTK*).[171] Studies reveal that when *SOX4* or *TNC* expression is reduced in a breast cancer cell line that is highly metastatic to the lung (LM2 cells), the cells exhibit morphology changes, have decreased motility, lower invasive potential and blocked lung metastasis *in vivo* when *SOX4* or *TNC* treated LM2 cells are injected into immunocompromised mice.[171] These results phenocopy those observed when miR-335 expression is restored in these cell lines. Although no targets have yet been identified for miR-126 linked to cellular proliferation, this miRNA functions as an important regulator of endothelial cell processes related to angiogenesis and vascular integrity both *in vitro* and *in vivo*. This is accomplished primarily by repressing the targets Sprouty-related *EVH1* domain-containing protein 1 (*SPRED1*) and the regulatory subunit of *PI3K*, *PIK3R2*, which in turn results in activating *VEGF* signaling.[172] Due to the strong correlation between loss of miR-335 and miR-126 expression with metastatic relapse, these two miRNAs hold great promise not only as a prognostic tool for breast cancer but also as therapeutic targets that control metastatic processes related to cell migration, modulation of the extracellular matrix and vascular remodeling.

6.6 Conclusions and Future Outlook

Accumulating experimental evidence supports the notion that miRNAs play a direct role in cancer progression by regulating targets that control processes related to cellular growth, differentiation, apoptosis, angiogenesis, and motility. There is also a clear intersection between miRNAs and many established tumor suppressor genes and oncogenes, e.g., p53, *Myc*, and *RAS*. Thus, non-coding RNAs and protein-coding genes exist in complex regulatory feedback loops to fine-tune expression thresholds and direct cellular proliferation rates and cell cycle check points. Tumor formation therefore occurs when this balance is tipped towards a growth state due to gene misexpression. Furthermore, miRNA profiling studies on a large range of human cancers reveals that each tissue and likely each diseased state possess their own miRNA expression 'signature' that will facilitate in the development of novel small RNA-based diagnostic and prognostic biomarkers. Although many of these miRNAs are likely mis-expressed as a secondary consequence of the disease, certain miRNAs such as those discussed in this chapter are clearly functioning as *bona fide* tumor suppressor genes, oncogenes, and metastatic factors. These cancer-related miRNAs are promising targets for small RNA-based therapies since they are found to regulate multiple targets residing in distinct genetic pathways that work together to direct cellular growth and differentiation, and thus have the ability to control tumorigenesis from many angles. Indeed, recent proof-of-principle experiments in mice have shown that systemic delivery of miRNA-expressing viruses, synthetic miRNA mimics and antisense miRNA inhibitors can modulate tumor formation without inducing toxicities outside of the intended organ. However, better understanding of the molecules involved in controlling miRNA promoter transcription, miRNA processing, and miRNA-mediated regulation as well as the development of organ- and cell-specific delivery systems will be crucial when using miRNAs in the clinic to treat cancer. Extensive characterization of the complete list of miRNA-target interactions will be imperative to understand the full range of effects a single miRNA can elicit, especially in relation to the cellular context and disease state. Clearly, certain miRNA genes such as miR-26a, the miR-29 family, and the miR-17-92 cluster regulate a multitude of targets that dictate either oncogenic or tumor suppressor roles in specific situations and thus will make miRNA-based therapies more challenging. Nonetheless, miRNAs have emerged as a major class of regulatory molecules that play essential roles during development and disease, and possess immense diagnostic and therapeutic potential.

References

1. Esquela-Kerscher A. and Slack F.J. (2006). Oncomirs — microRNAs with a role in cancer. *Nat Rev Cancer* **6**, 259–269.
2. Calin G.A. and Croce C.M. (2006). MicroRNA signatures in human cancers. *Nat Rev Cancer* **6**, 857–866.
3. Griffiths-Jones S., Saini H.K., van Dongen S. *et al.* (2008). MiRBase: Tools for microRNA genomics. *Nucleic Acids Res* **36** (Database issue), D154.
4. Stefani G. and Slack F.J. (2008). Small non-coding RNAs in animal development. *Nat Rev Mol Cell Biol* **9**, 219–230.
5. Reinhart B., Slack F.J., Basson M. *et al.* (2000). The 21 nucleotide *let-7* RNA regulates *C. elegans* developmental timing. *Nature* **403**, 901–906.
6. Chalfie M., Horvitz H.R. and Sulston J.E. (1981). Mutations that lead to reiterations in the cell lineages of *C. elegans*. *Cell* **24**, 59–69.
7. Moss E.G. (2007). Heterochronic genes and the nature of developmental time. *Curr Biol* **17**, R425–434.
8. Sempere L.F., Sokol N.S., Dubrovsky E.B. *et al.* (2003). Temporal regulation of microRNA expression in *Drosophila melanogaster* mediated by hormonal signals and *broad-Complex* gene activity. *Dev Biol* **259**, 9–18.
9. Bashirullah A., Pasquinelli A.E., Kiger A.A. *et al.* (2003). Coordinate regulation of small temporal RNAs at the onset of Drosophila metamorphosis. *Dev Biol* **259**, 1–8.

10. Caygill E.E. and Johnston L.A. (2008). Temporal regulation of metamorphic processes in Drosophila by the *let-7* and miR-125 heterochronic microRNAs. *Curr Biol* **18**, 943–950.

11. Sokol N.S., Xu P., Jan Y.N. *et al.* (2008). Drosophila *let-7* microRNA is required for remodeling of the neuromusculature during metamorphosis. *Genes Dev* **22**, 1591–1596.

12. Thomson J.M., Newman M., Parker J.S. *et al.* (2006). Extensive post-transcriptional regulation of microRNAs and its implications for cancer. *Genes Dev* **20**, 2202–2207.

13. Wienholds E. and Plasterk R.H. (2005). MicroRNA function in animal development. *FEBS Lett* **579**, 5911–5922.

14. Wulczyn F.G., Smirnova L., Rybak A. *et al.* (2007). Post-transcriptional regulation of the *let-7* microRNA during neural cell specification. *FASEB J* **21**, 415–426.

15. Rybak A., Fuchs H., Smirnova L. *et al.* (2008). A feedback loop comprising lin-28 and *let-7* controls pre-*let-7* maturation during neural stem-cell commitment. *Nat Cell Biol* **10**, 987–993.

16. Schulman B.R., Esquela-Kerscher A. and Slack F.J. (2005). Reciprocal expression of lin-41 and the microRNAs *let-7* and mir-125 during mouse embryogenesis. *Dev Dyn* **234**, 1046–1054.

17. Lee Y.S., Kim H.K., Chung S. *et al.* (2005). Depletion of human micro-RNA miR-125b reveals that it is critical for the proliferation of differentiated cells but not for the down-regulation of putative targets during differentiation. *J Biol Chem* **280**, 16635–16641.

18. Mansfield J.H., Harfe B.D., Nissen R. *et al.* (2004). MicroRNA-responsive 'sensor' transgenes uncover Hox-like and other developmentally regulated patterns of vertebrate microRNA expression. *Nat Genet* **36**, 1079–1083.

19. Lagos-Quintana M., Rauhut R., Yalcin A. *et al.* (2002). Identification of tissue-specific microRNAs from mouse. *Curr Biol* **12**, 735–739.

20. Sempere L.F., Freemantle S., Pitha-Rowe I. *et al.* (2004). Expression profiling of mammalian microRNAs uncovers a subset of brain-expressed microRNAs with possible roles in murine and human neuronal differentiation. *Genome Biol* **5**, R13.1–11.

21. Wu L. and Belasco J.G. (2005). Micro-RNA regulation of the mammalian lin-28 gene during neuronal differentiation of embryonal carcinoma cells. *Mol Cell Biol* **25**, 9198–9208.

22. Johnson S.M., Grosshans H., Shingara J. *et al.* (2005). RAS is regulated by the *let-7* microRNA family. *Cell* **120**, 635–647.

23. Calin G.A., Sevignani C., Dumitru C.D. *et al.* (2004). Human microRNA genes are frequently located at fragile sites and genomic regions involved in cancers. *Proc Natl Acad Sci USA* **101**, 2999–3004.

24. Iorio M.V., Ferracin M., Liu C.G. *et al.* (2005). MicroRNA gene expression deregulation in human breast cancer. *Cancer Res* **65**, 7065–7070.

25. Sonoki T., Iwanaga E., Mitsuya H. *et al.* (2005). Insertion of microRNA-125b-1, a human homologue of lin-4, into a rearranged immunoglobulin heavy chain gene locus in a patient with precursor B-cell acute lymphoblastic leukemia. *Leukemia* **11**, 2009–2010.

26. Johnson C.D., Esquela-Kerscher A., Stefani G. *et al.* (2007). The *let-7* microRNA represses cell proliferation pathways in human cells. *Cancer Res* **67**, 7713–7722.

27. Esquela-Kerscher A., Trang P., Wiggins J.F. *et al.* (2008). The *let-7* microRNA reduces tumor growth in mouse models of lung cancer. *Cell Cycle* **7**, 759–764.

28. Takamizawa J., Konishi H., Yanagisawa K. *et al.* (2004). Reduced expression of the *let-7* microRNAs in human lung cancers in association with shortened postoperative survival. *Cancer Res* **64**, 3753–3756.

29. Yu F., Yao H., Zhu P. *et al.* (2007). *let-7* regulates self renewal and tumorigenicity of breast cancer cells. *Cell* **131**, 1109–1123.

30. Lee Y.S. and Dutta A. (2007). The tumor suppressor microRNA *let-7* represses the HMGA2 oncogene. *Genes Dev* **21**, 1025–1030.

31. Kumar M.S., Erkeland, S.J., Pester R.E. *et al.* (2008). Suppression of non-small cell lung tumor development by the *let-7* microRNA family. *Proc Natl Acad Sci USA* **105**, 3903–3908.

32. Yu S.L., Chen H.Y., Chang G.C. *et al.* (2008). MicroRNA signature predicts survival and relapse in lung cancer. *Cancer Cell* **13**, 48–57.

33. Brennecke J., Hipfner D.R., Stark A. *et al.* (2003). *Bantam* encodes a developmentally regulated microRNA that controls cell proliferation and regulates the proapoptotic gene *hid* in *Drosophila*. *Cell* **113**, 25–36.

34. Hipfner D.R., Weigmann K. and Cohen S.M. (2002). The bantam gene regulates Drosophila growth. *Genetics* **161**, 1527–1537.

35. Xu P., Vernooy S.Y., Guo M. *et al.* (2003). The *Drosophila* microRNA *mir-14* suppresses cell death and is required for normal fat metabolism. *Curr Biol* **13**, 790–795.

36. Stark A., Brennecke J., Russell R.B. *et al.* (2003). Identification of *Drosophila* microRNA targets. *PLoS Biol* **1**, 397–409.

37. Leaman D., Chen P.Y., Fak J. *et al* (2005). Antisense-mediated depletion reveals essential and specific functions of microRNAs in Drosophila development. *Cell* **121**, 1097–1108.

38. Thompson B.J. and Cohen S.M. (2006). The Hippo pathway regulates the bantam microRNA to control cell proliferation and apoptosis in Drosophila. *Cell* **126**, 767–774.

39. Johnston R.J. and Hobert O. (2003). A microRNA controlling left/right neuronal asymmetry in *Caenorhabditis elegans*. *Nature* **426**, 845–849.

40. Chang S., Johnston Jr. R.J., Frokjaer-Jensen C. *et al.* (2004). MicroRNAs act sequentially and asymmetrically to control chemosensory laterality in the nematode. *Nature* **430**, 785–789.

41. Hobert O. (2006). Architecture of a microRNA-controlled gene regulatory network that diversifies neuronal cell fates. *Cold Spring Harb Symp Quant Biol* **71**, 181–188.

42. Giraldez A.J., Cinalli R.M., Glasner M.E. *et al.* (2005). MicroRNAs regulate brain morphogenesis in zebrafish. *Science* **308**, 833–838.

43. Poy M.N., Eliasson L., Krutzfeldt J. *et al.* (2004). A pancreatic islet-specific microRNA regulates insulin secretion. *Nature* **432**, 226–230.

44. Esau C., Kang X., Peralta E. *et al.* (2004). MicroRNA-143 regulates adipocyte differentiation. *J Biol Chem* **279**, 52361–52365.

45. Hornstein E., Mansfield J.H., Yekta S. *et al.* (2005). The microRNA miR-196 acts upstream of Hoxb8 and Shh in limb development. *Nature* **438**, 671–674.

46. Rodriguez A., Vigorito E., Clare S. *et al.* (2007). Requirement of bic/microRNA-155 for normal immune function. *Science* **316**, 608–611.

47. Thai T.H., Calado D.P., Casola S. *et al.* (2007). Regulation of the germinal center response by microRNA-155. *Science* **316**, 604–608.

48. Zhao Y., Samal E. and Srivastava D. (2005). Serum response factor regulates a muscle-specific microRNA that targets Hand2 during cardiogenesis. *Nature* **436**, 214–220.

49. Zhao Y., Ransom J.F., Li A. *et al.* (2007). Dysregulation of cardiogenesis, cardiac conduction, and cell cycle in mice lacking miRNA-1-2. *Cell* **129**, 303–317.

50. Chen J.F., Mandel E.M., Thomson J.M. *et al.* (2006). The role of microRNA-1 and microRNA-133 in skeletal muscle proliferation and differentiation. *Nat Genet* **38**, 228–233.

51. van Rooij E., Sutherland L.B., Qi X. *et al.* (2007). Control of stress-dependent cardiac growth and gene expression by a microRNA. *Science* **316**, 575–579.

52. van Rooij E., Sutherland L.B., Liu N. *et al.* (2006). A signature pattern of stress-responsive microRNAs that can evoke cardiac hypertrophy and heart failure. *Proc Natl Acad Sci USA* **103**, 18255–18260.

53. Li Q.J., Chau J., Ebert P.J. *et al.* (2007). miR-181a is an intrinsic modulator of T cell sensitivity and selection. *Cell* **129**, 147–161.

54. Chen C.Z., Li L., Lodish H.F. *et al.* (2004). MicroRNAs modulate hematopoietic lineage differentiation. *Science* **303**, 83–86.

55. Naguibneva I., Ameyar-Zazoua M., Polesskaya A. *et al.* (2006). The microRNA miR-181 targets the homeobox protein Hox-A11 during mammalian myoblast differentiation. *Nat Cell Biol* **8**, 278–284.

56. Chan S.P. and Slack F.J. (2007). And now introducing mammalian mirtrons. *Dev Cell* **13**, 605–607.

57. Filipowicz W., Bhattacharyya S.N. and Sonenberg N. (2008). Mechanisms of post-transcriptional regulation by microRNAs: Are the answers in sight? *Nat Rev Genet* **9**, 102–114.

58. O'Donnell K.A., Wentzel E.A., Zeller K.I. *et al.* (2005). c-Myc-regulated microRNAs modulate E2F1 expression. *Nature* **435**, 839–843.

59. Chang T.C., Yu D., Lee Y.S. *et al.* (2008). Widespread microRNA repression by Myc contributes to tumorigenesis. *Nat Genet* **40**, 43–50.

60. He L., He X., Lim L.P. *et al.* (2007). A microRNA component of the p53 tumor suppressor network. *Nature* **447**, 1130–1134.

61. Chang T.C., Wentzel E.A., Kent O.A. *et al.* (2007). Transactivation of miR-34a by p53 broadly influences gene expression and promotes apoptosis. *Mol Cell* **26**, 745–752.

62. Bommer G.T., Gerin I., Feng Y. *et al.* (2007). p53-mediated activation of miRNA34 candidate tumor-suppressor genes. *Curr Biol* **17**, 1298–1307.

63. Raver-Shapira N., Marciano E., Meiri E. *et al.* (2007). Transcriptional activation of miR-34a contributes to p53-mediated apoptosis. *Mol Cell* **26**, 731–743.

64. Tarasov V., Jung P., Verdoodt B. *et al.* (2007). Differential regulation of microRNAs by p53 revealed by massively parallel sequencing: miR-34a is a p53 target that induces apoptosis and G1-arrest. *Cell Cycle* **6**, 1586–1593.

65. Welch C., Chen Y. and Stallings R.L. (2007). MicroRNA-34a functions as a potential tumor suppressor by inducing apoptosis in neuroblastoma cells. *Oncogene* **26**, 5017–5022.

66. Tazawa H., Tsuchiya N., Izumiya M. *et al.* (2007). Tumor-suppressive miR-34a induces senescence-like growth arrest through modulation of the E2F pathway in human colon cancer cells. *Proc Natl Acad Sci USA* **104**, 15472–15477.

67. Cole K.A., Attiyeh E.F., Mosse Y.P. *et al.* (2008). A functional screen identifies miR-34a as a candidate neuroblastoma tumor suppressor gene. *Mol Cancer Res* **6**, 735–742.

68. Wei J.S., Song Y.K., Durinck S. *et al.* (2008). The MYCN oncogene is a direct target of miR-34a. *Oncogene* **27**, 5204–5213.

69. Saito Y., Liang G., Egger G. *et al.* (2006). Specific activation of microRNA-127 with downregulation of the proto-oncogene BCL6 by chromatin-modifying drugs in human cancer cells. *Cancer Cell* **9**, 435–443.
70. Lujambio A., Ropero S., Ballestar E. *et al.* (2007). Genetic unmasking of an epigenetically silenced microRNA in human cancer cells. *Cancer Res* **67**, 1424–1429.
71. Brueckner B., Stresemann C., Kuner R. *et al.* (2007). The human *let-7a-3* locus contains an epigenetically regulated microRNA gene with oncogenic function. *Cancer Res* **67**, 1419–1423.
72. Kozaki K., Imoto I., Mogi S. *et al.* (2008). Exploration of tumor-suppressive microRNAs silenced by DNA hypermethylation in oral cancer. *Cancer Res* **68**, 2094–2105.
73. Toyota M., Suzuki H., Sasaki Y. *et al.* (2008). Epigenetic silencing of microRNA-34b/c and B-cell translocation gene 4 is associated with CpG island methylation in colorectal cancer. *Cancer Res* **68**, 4123–4132.
74. Han L., Witmer P.D., Casey E. *et al.* (2007). DNA methylation regulates MicroRNA expression. *Cancer Biol Ther* **6**, 1284–1288.
75. Lujambio A., Calin G.A., Villanueva A. *et al.* (2008). A microRNA DNA methylation signature for human cancer metastasis. *Proc Natl Acad Sci USA* **105**, 13556–13561.
76. Ji Q., Hao X., Meng Y. *et al.* (2008). Restoration of tumor suppressor miR-34 inhibits human p53-mutant gastric cancer tumorspheres. *BMC Cancer* **8**, 266–278.
77. Rokhlin O.W., Scheinker V.S., Taghiyev A.F. *et al.* (2008). MicroRNA-34 mediates AR-dependent p53-induced apoptosis in prostate cancer. *Cancer Biol Ther* **7**, 1288–1296.
78. Karube Y., Tanaka H., Osada H. *et al.* (2005). Reduced expression of Dicer associated with poor prognosis in lung cancer patients. *Cancer Sci* **96**, 111–115.
79. Merritt W.M., Lin Y.G., Han L.Y. *et al.* (2008). Dicer, Drosha, and outcomes in patients with ovarian cancer. *N Engl J Med* **359**, 2641–2650.
80. Kumar M.S., Lu J., Mercer K.L. *et al.* (2007). Impaired microRNA processing enhances cellular transformation and tumorigenesis. *Nat Genet* **39**, 673–677.
81. Chendrimada T.P., Gregory R.I., Kumaraswamy E. *et al.* (2005). TRBP recruits the Dicer complex to Ago2 for microRNA processing and gene silencing. *Nature* **436**, 740–744.
82. Melo S.A., Ropero S., Moutinho C. *et al.* (2009). A TARBP2 mutation in human cancer impairs microRNA processing and DICER1 function. *Nat Genet* **41**, 365–370.

83. Newman M.A., Thomson J.M. and Hammond S.M. (2008). Lin-28 interaction with the *Let-7* precursor loop mediates regulated microRNA processing. *RNA* **14**, 1539–1549.
84. Viswanathan S.R., Daley G.Q. and Gregory R.I. (2008). Selective blockade of microRNA processing by Lin28. *Science* **320**, 97–100.
85. Piskounova E., Viswanathan S.R., Janas M. *et al.* (2008). Determinants of microRNA processing inhibition by the developmentally regulated RNA-binding protein Lin28. *J Biol Chem* **283**, 21310–21314.
86. Heo I., Joo C., Cho J. *et al.* (2008). Lin28 mediates the terminal uridylation of *let-7* precursor MicroRNA. *Mol Cell* **32**, 276–284.
87. Yu J., Vodyanik M.A., Smuga-Otto K. *et al.* (2007). Induced pluripotent stem cell lines derived from human somatic cells. *Science* **318**, 1917–1920.
88. West J.A., Viswanathan S.R., Yabuuchi A. *et al.* (2009). A role for Lin28 in primordial germ-cell development and germ-cell malignancy. *Nature* **460**, 909–913.
89. Guo Y., Chen Y., Ito H. *et al.* (2006). Identification and characterization of lin-28 homolog B (LIN28B) in human hepatocellular carcinoma. *Gene* **384**, 51–61.
90. Carmell M.A., Xuan Z., Zhang M.Q. *et al.* (2002). The Argonaute family: Tentacles that reach into RNAi, developmental control, stem cell maintenance, and tumorigenesis. *Genes Dev* **16**, 2733–2742.
91. Nelson P., Kiriakidou M., Sharma A. *et al.* (2003). The microRNA world: Small is mighty. *Trends Biochem Sci* **28**, 534–540.
92. Adams B.D., Claffey K.P. and White B.A. (2009). Argonaute-2 expression is regulated by epidermal growth factor receptor and mitogen-activated protein kinase signaling and correlates with a transformed phenotype in breast cancer cells. *Endocrinology* **150**, 14–23.
93. Yanaihara N., Caplen N., Bowman E. *et al.* (2006). Unique microRNA molecular profiles in lung cancer diagnosis and prognosis. *Cancer Cell* **9**, 189–198.
94. Lu J., Getz G., Miska E.A. *et al.* (2005). MicroRNA expression profiles classify human cancers. *Nature* **435**, 834–838.
95. Gaur A., Jewell D.A., Liang Y. *et al.* (2007). Characterization of microRNA expression levels and their biological correlates in human cancer cell lines. *Cancer Res* **67**, 2456–2468.
96. Michael M.Z., O'Connor S.M., van Holst Pellekaan N.G. *et al.* (2003). Reduced accumulation of specific microRNAs in colorectal neoplasia. *Mol Cancer Res* **1**, 882–891.

97. Calin G.A., Liu C.G., Sevignani C. *et al.* (2004). MicroRNA profiling reveals distinct signatures in B cell chronic lymphocytic leukemias. *Proc Natl Acad Sci USA* **101**, 11755–11760.

98. He H., Jazdzewski K., Li W. *et al.* (2005). The role of microRNA genes in papillary thyroid carcinoma. *Proc Natl Acad Sci USA* **102**, 19075–19080.

99. Volinia S., Calin G.A., Liu C.G. *et al.* (2006). A microRNA expression signature of human solid tumors defines cancer gene targets. *Proc Natl Acad Sci USA* **103**, 2257–2261.

100. Kent O.A. and Mendell J.T. (2006). A small piece in the cancer puzzle: MicroRNAs as tumor suppressors and oncogenes. *Oncogene* **25**, 6188–6196.

101. Calin G.A., Dumitru C.D., Shimizu M. *et al.*(2002). Frequent deletions and downregulation of microRNA genes *miR15* and *miR16* at 13q14 in chronic lymphocytic leukemia. *Proc Natl Acad Sci USA* **99**, 15524–15529.

102. Calin G.A., Ferracin M., Cimmino A. *et al.* (2005). A microRNA signature associated with prognosis and progression in chronic lymphocytic leukemia. *N Engl J Med* **353**, 1793–1801.

103. Cimmino A., Calin G.A., Fabbri M. *et al.* (2005). MiR-15 and miR-16 induce apoptosis by targeting BCL2. *Proc Natl Acad Sci USA* **102**,13944–13949.

104. Calin G.A., Cimmino A., Fabbri M. *et al.* (2008). MiR-15a and miR-16-1 cluster functions in human leukemia. *Proc Natl Acad Sci USA* **105**, 5166–5171.

105 Linsley P.S., Schelter J., Burchard J. *et al.* (2007). Transcripts targeted by the microRNA-16 family cooperatively regulate cell cycle progression. *Mol Cell Biol* **27**, 2240–2252.

106. Bonci D., Coppola V., Musumeci M. *et al.* (2008). The miR-15a-miR-16-1 cluster controls prostate cancer by targeting multiple oncogenic activities. *Nat Med* **14**, 1271–1277.

107. Grosshans H., Johnson T., Reinert K.L. *et al.* (2005). The temporal patterning microRNA *let-7* regulates several transcription factors at the larval to adult transition in *C. elegans. Dev Cell* **8**, 321–330.

108. Mayr C., Hemann M.T. and Bartel D.P. (2007). Disrupting the pairing between *let-7* and Hmga2 enhances oncogenic transformation. *Science* **315**, 1576–1579.

109. Xi Y., Shalgi R., Fodstad O. *et al.* (2006). Differentially regulated micro-RNAs and actively translated messenger RNA transcripts by tumor suppressor p53 in colon cancer. *Clin Cancer Res* **12**, 2014–2024.

110. Visone R., Pallante P., Vecchione A. *et al.* (2007). Specific microRNAs are downregulated in human thyroid anaplastic carcinomas. *Oncogene* **26**, 7590–7595.

111. Kota J., Chivukula R.R., O'Donnell K.A. *et al.* (2009). Therapeutic microRNA delivery suppresses tumorigenesis in a murine liver cancer model. *Cell* **137**, 1005–1017.

112. Kashuba V.I., Li J., Wang F. *et al.* (2004). RBSP3 (HYA22) is a tumor suppressor gene implicated in major epithelial malignancies. *Proc Natl Acad Sci USA* **101**, 4906–4911.

113. Huse J.T., Brennan C., Hambardzumyan D. *et al.* (2009). The PTEN-regulating microRNA miR-26a is amplified in high-grade glioma and facilitates gliomagenesis *in vivo. Genes Dev* **23**, 1327–1337.

114. Cummins J.M., He Y., Leary R.J. *et al.* (2006). The colorectal microRNAome. *Proc Natl Acad Sci USA* **103**, 3687–3692.

115. Mott J.L., Kobayashi S., Bronk S.F. *et al.* (2007). mir-29 regulates Mcl-1 protein expression and apoptosis. *Oncogene* **26**, 6133–6140.

116. Park S.Y., Lee J.H., Ha M. *et al.* (2009). miR-29 miRNAs activate p53 by targeting p85 alpha and CDC42. *Nat Struct Mol Biol* **16**, 23–29.

117. Fabbri M., Garzon R., Cimmino A. *et al.* (2007). MicroRNA-29 family reverts aberrant methylation in lung cancer by targeting DNA methyltransferases 3A and 3B. *Proc Natl Acad Sci USA* **104**, 15805–15810.

118. Pekarsky Y., Santanam U., Cimmino A. *et al.* (2006). Tcl1 expression in chronic lymphocytic leukemia is regulated by miR-29 and miR-181. *Cancer Res* **66**, 11590–11593.

119. Wang H., Garzon R., Sun H. *et al.* (2008). NF-kappaB-YY1-miR-29 regulatory circuitry in skeletal myogenesis and rhabdomyosarcoma. *Cancer Cell* **14**, 369–381.

120. Gebeshuber C.A., Zatloukal K. and Martinez J. (2009). miR-29a suppresses tristetraprolin, which is a regulator of epithelial polarity and metastasis. *EMBO Rep* **10**, 400–405.

121. Metzler M., Wilda M., Busch K. *et al.* (2004). High expression of precursor microRNA-155/BIC RNA in children with Burkitt lymphoma. *Genes Chromosomes Cancer* **39**, 167–169.

122. Clurman B.E. and Hayward W.S. (1989). Multiple proto-oncogene activations in avian leukosis virus-induced lymphomas: Evidence for stage-specific events. *Mol Cell Biol* **9**, 2657–2664.

123. Tam W., Ben-Yehuda D. and Hayward W.S. (1997). bic, a novel gene activated by proviral insertions in avian leukosis virus-induced lymphomas, is likely to function through its noncoding RNA. *Mol Cell Biol* **17**, 1490–1502.

124. Eis P.S., Tam W., Sun L. *et al.* (2005). Accumulation of miR-155 and BIC RNA in human B cell lymphomas. *Proc Natl Acad Sci USA* **102**, 3627–3632.

125. Kluiver J., Poppema S., de Jong D. *et al.* (2005). BIC and miR-155 are highly expressed in Hodgkin, primary mediastinal and diffuse large B cell lymphomas. *J Pathol* **207**, 243–249.

126. van den Berg, A., Kroesen B.J., Kooistra K. *et al.* (2003). High expression of B-cell receptor inducible gene BIC in all subtypes of Hodgkin lymphoma. *Gene Chromosome Canc* **37**, 20–28.

127. Lee E.J., Gusev Y., Jiang J. *et al.* (2007). Expression profiling identifies microRNA signature in pancreatic cancer. *Int J Cancer* **120**, 1046–1054.

128. Szafranska A.E., Davison T.S., John J. *et al.* (2007). MicroRNA expression alterations are linked to tumorigenesis and non-neoplastic processes in pancreatic ductal adenocarcinoma. *Oncogene* **26**, 4442–4452.

129. Costinean S., Zanesi N., Pekarsky Y. *et al.* (2006). Pre-B cell proliferation and lymphoblastic leukemia/high-grade lymphoma in E(mu)-miR155 transgenic mice. *Proc Natl Acad Sci USA* **103**, 7024–7029.

130. Skalsky R.L., Samols M.A., Plaisance K.B. *et al.* (2007). Kaposi's sarcoma-associated herpesvirus encodes an ortholog of miR-155. *J Virol* **81**, 12836–12845.

131. Gottwein E., Mukherjee N., Sachse C. *et al.* (2007). A viral microRNA functions as an orthologue of cellular miR-155. *Nature* **450**, 1096–1099.

132. Dorsett Y., McBride K.M., Jankovic M. *et al.* (2008). MicroRNA-155 suppresses activation-induced cytidine deaminase-mediated Myc-Igh translocation. *Immunity* **28**, 630–638.

133. Teng G., Hakimpour P., Landgraf P. *et al.* (2008). MicroRNA-155 is a negative regulator of activation-induced cytidine deaminase. *Immunity* **28**, 621–629.

134. Ota A., Tagawa H., Karnan S. *et al.* (2004). Identification and characterization of a novel gene, C13orf25, as a target for 13q31-q32 amplification in malignant lymphoma. *Cancer Res* **64**, 3087–3095.

135. Gordon A.T., Brinkschmidt C., Anderson J. *et al.* (2000). A novel and consistent amplicon at 13q31 associated with alveolar rhabdomyosarcoma. *Gene Chromosome Canc* **28**, 220–226.

136. Schmidt H., Bartel F., Kappler M. *et al.* (2005). Gains of 13q are correlated with a poor prognosis in liposarcoma. *Mod Pathol* **18**, 638–644.

137. Petrocca F., Visone R., Onelli M.R. *et al.* (2008). E2F1-regulated microRNAs impair TGFbeta-dependent cell-cycle arrest and apoptosis in gastric cancer. *Cancer Cell* **13**, 272–286.

138. Tagawa H. and Seto M. (2005). A microRNA cluster as a target of genomic amplification in malignant lymphoma. *Leukemia* **19**, 2013–2016.

139. Hayashita Y., Osada H., Tatematsu Y. *et al.* (2005). A polycistronic microRNA cluster, miR-17-92, is overexpressed in human lung cancers and enhances cell proliferation. *Cancer Res* **65**, 9628–9632.

140. Cui J.W., Li Y.J., Sarkar A. *et al.* (2007). Retroviral insertional activation of the Fli-3 locus in erythroleukemias encoding a cluster of microRNAs that convert Epo-induced differentiation to proliferation. *Blood* **110**, 2631–2640.

141. Wang C.L., Wang B.B., Bartha G. *et al.* Activation of an oncogenic microRNA cistron by provirus integration. *Proc Natl Acad Sci USA* **103**, 18680–18684.

142. Ventura A., Young A.G., Winslow M.M. *et al.* (2008). Targeted deletion reveals essential and overlapping functions of the miR-17 through 92 family of miRNA clusters. *Cell* **132**, 875–886.

143. He L., Thomson J.M., Hemann M.T. *et al.* (2005). A microRNA polycistron as a potential human oncogene. *Nature* **435**, 828–833.

144. Xiao C., Srinivasan L., Calado D.P. *et al.* (2008). Lymphoproliferative disease and autoimmunity in mice with increased miR-17-92 expression in lymphocytes. *Nat Immunol* **9**, 405–414.

145. Trimarchi J.M. and Lees J.A. (2002). Sibling rivalry in the E2F family. *Nat Rev Mol Cell Biol* **3**, 11–20.

146. Sylvestre Y., De Guire V., Querido E. *et al.* (2007). An E2F/miR-20a autoregulatory feedback loop. *J Biol Chem* **282**, 2135–2143.

147. Woods K., Thomson J.M. and Hammond S.M. (2007). Direct regulation of an oncogenic micro-RNA cluster by E2F transcription factors. *J Biol Chem* **282**, 2130–2134.

148. Ivanovska I., Ball A.S., Diaz R.L. *et al.* (2008). MicroRNAs in the miR-106b family regulate p21/CDKN1A and promote cell cycle progression. *Mol Cell Biol* **28**, 2167–2174.

149. Koralov S.B., Muljo S.A., Galler G.R. *et al.* (2008). Dicer ablation affects antibody diversity and cell survival in the B lymphocyte lineage. *Cell* **132**, 860–874.

150. Dews M., Homayouni A., Yu D. *et al.* (2006). Augmentation of tumor angiogenesis by a Myc-activated microRNA cluster. *Nat Genet* **38**, 1060–1065.

151. Lin Y.W., Sheu J.C., Liu L.Y. *et al.* (1999). Loss of heterozygosity at chromosome 13q in hepatocellular carcinoma: Identification of three independent regions. *Eur J Cancer* **35**, 1730–1734.

152. Zhang L., Huang J., Yang N. *et al.* (2006). MicroRNAs exhibit high frequency genomic alterations in human cancer. *Proc Natl Acad Sci USA* **103**, 9136–9141.

153. Hossain A., Kuo M.T. and Saunders G.F. (2006). Mir-17-5p regulates breast cancer cell proliferation by inhibiting translation of AIB1 mRNA. *Mol Cell Biol* **26**, 8191–8201.

154. Landgraf P., Rusu M., Sheridan R. *et al.* (2007). A mammalian microRNA expression atlas based on small RNA library sequencing. *Cell* **129**, 1401–1414.

155. Chan J.A., Krichevsky A.M. and Kosik K.S. (2005). MicroRNA-21 is an anti-apoptotic factor in human glioblastoma cells. *Cancer Res* **65**, 6029–6033.

156. Ciafre S.A., Galardi S., Mangiola A. *et al.* (2005). Extensive modulation of a set of microRNAs in primary glioblastoma. *Biochem Biophys Res Commun* **334**, 1351–1358.

157. Li T., Li D., Sha J. *et al.*(2009). MicroRNA-21 directly targets MARCKS and promotes apoptosis resistance and invasion in prostate cancer cells. *Biochem Biophys Res Commun* **383**, 280–285.

158. Si M.L., Zhu S., Wu H. *et al.* (2007). miR-21-mediated tumor growth. *Oncogene* **26**, 2799–2803.

159. Zhu S., Wu H., Wu F. *et al.* (2008). MicroRNA-21 targets tumor suppressor genes in invasion and metastasis. *Cell Res* **18**, 350–359.

160. Asangani I.A., Rasheed S.A., Nikolova D.A. *et al.* (2008). MicroRNA-21 (miR-21) post-transcriptionally downregulates tumor suppressor Pdcd4 and stimulates invasion, intravasation and metastasis in colorectal cancer. *Oncogene* **27**, 2128–2136.

161. Gabriely G., Wurdinger T., Kesari S. *et al.* (2008). MicroRNA 21 promotes glioma invasion by targeting matrix metalloproteinase regulators. *Mol Cell Biol* **28**, 5369–5380.

162. Sayed D., Rane S., Lypowy J. *et al.* (2008). MicroRNA-21 targets Sprouty2 and promotes cellular outgrowths. *Mol Biol Cell* **19**, 3272–3282.

163. Krichevsky A.M. and Gabriely G. (2009). miR-21: A small multi-faceted RNA. *J Cell Mol Med* **13**, 39–53.

164. Ma L., Teruya-Feldstein J. and Weinberg R.A. (2007). Tumor invasion and metastasis initiated by microRNA-10b in breast cancer. *Nature* **449**, 682–688.

165. Huang Q., Gumireddy K., Schrier M. *et al.* (2008). The microRNAs miR-373 and miR-520c promote tumor invasion and metastasis. *Nat Cell Biol* **10**, 202–210.

166. Pereira P.A., Rubenthiran U., Kaneko M. *et al.* (2001). CD44s expression mitigates the phenotype of human colorectal cancer hepatic metastases. *Anticancer Res* **21**, 2713–2717.

167. Gong Y., Sun X., Huo L. *et al.* (2005). Expression of cell adhesion molecules, CD44s and E-cadherin, and microvessel density in invasive micropapillary carcinoma of the breast. *Histopathology* **46**, 24–30.

168. Jaeger E.B., Samant R.S. and Rinker-Schaeffer C.W. (2001). Metastasis suppression in prostate cancer. *Cancer Metastasis Rev* **20**, 279–286.

169. Choi S.H., Takahashi K., Eto H. *et al.* (2000). CD44s expression in human colon carcinomas influences growth of liver metastases. *Int J Cancer* **85**, 523–526.
170. Voorhoeve P.M., le Sage C., Schrier M. *et al.* (2006). A genetic screen implicates miRNA-372 and miRNA-373 as oncogenes in testicular germ cell tumors. *Cell* **124**, 1169–1181.
171. Tavazoie S.F., Alarcon C., Oskarsson T. *et al.* (2008). Endogenous human microRNAs that suppress breast cancer metastasis. *Nature* **451**, 147–152.
172. Fish J.E., Santoro M.M., Morton S.U. *et al.* (2008). miR-126 regulates angiogenic signaling and vascular integrity. *Dev Cell* **15**, 272–284.

7

Simultaneous Detection of Primary, Precursor and Mature MicroRNAs by qPCR

Jinmai Jiang, Eun Joo Lee* and Thomas D. Schmittgen*,†*

MicroRNA (miRNA) is transcribed as a long primary precursor transcript (pri-miRNA) that is subsequently processed in the nucleus by Drosha/DGCR8 to produce the precursor miRNA (pre-miRNA), a stable hairpin of approximately 70 nts in length. The pre-miRNA is processed by Dicer in the cytoplasm to produce the approximately 21 nt mature miRNA. Numerous examples in the literature exist that describe the regulation of miRNA processing at either the Drosha or Dicer steps. By simultaneously quantifying the pri-, pre- and mature miRNA, it is possible to determine precisely at which step in the miRNA biogenesis regulation occurs. We present here a real-time quantitative RT-PCR assay that distinguishes the pri-, pre- and mature miRNA. This approach may be used to study transcriptional or post-transcriptional regulation of miRNA processing.

7.1 Introduction

MicroRNAs (miRNAs) are small, regulatory, non-coding RNAs. miRNA was discovered in *C. elegans* in 1993 by the Ambros and Ruvkun labs.[1,2]

---

*College of Pharmacy, Ohio State University, Columbus, OH 43210, USA. †E-mail: schmittgen.2@osu.edu.

The founding members of this class of RNAs include *lin-4* and *let-7*. In 2001, interest in the field peaked when hundreds of miRNAs were discovered in *Drosophila, C. elegans* and mammals.[3-5] Numerous miRNAs have been discovered in many species including 678 in humans, 472 in mice and 154 in *C. elegans*.[6] It has been estimated that miRNA regulates up to one third of human genes.[7]

MiRNA genes are located within introns, exons and in intergenic regions.[8] Although some exceptions exist, miRNAs are transcribed by RNA polymerase II.[9] The resulting miRNA transcript is referred to as the primary miRNA (pri-miRNA). The pri-miRNA is processed in the nucleus by a complex that includes the RNase III enzyme Drosha and the double-stranded RNA binding domain protein DGCR8.[10] The product of the Drosha processing is called the precursor miRNA (pre-miRNA). Both the pri- and pre-miRNA contain a stable stem-loop structure. The pre-miRNA is transported from the nucleus to the cytoplasm by the Ran-GTP-dependent Exportin 5.[11] Once inside the cytoplasm, the pre-miRNA is subsequently loaded into a complex that includes the RNase III enzyme Dicer and TRBP/Loquacious. This complex cleaves the loop from the pre-miRNA to produce a double-stranded complex composed of the miRNA and miRNA*. The miRNA* is believed to be the inactive strand and is typically degraded. The mature miRNA strand, along with the Argonaute protein Ago 2, is further assembled into a ribonucleoprotein complex known as RISC (RNA-induced silencing complex). RISC will then scan cellular RNAs until it locates the miRNA's target, which typically resides in the 3′ untranslated region (UTR) of messenger RNAs. The net result of the miRNA/messenger RNA interaction is a reduction in translation.

Conceivably, the expression of miRNA may be regulated at any point in its biogenesis. This includes at the level of transcription, due to gene duplication,[12] gene deletion[13] or promoter hypermethylation.[14,15] MiRNA biogenesis may also be regulated by Drosha processing, nuclear transport, subcellular localization,[16] terminal uridylation[17] or Dicer processing. Single nucleotide polymorphisms in the pri-miRNA caused a reduction in both pre- and mature miRNAs.[18,19] Several examples exist of miRNA biogenesis being regulated at the level of Drosha processing. Let-7 processing was inhibited by interaction of the RNA binding protein Lin 28 to the loop region of the pri-miRNA.[20,21] The post-transcriptional regulation of miRNA was studied in the developing mouse and it was shown that many miRNAs are transcribed at the precursor level. However, upon differentiation, the ubiquitously expressed precursor miRNAs are processed to mature miRNA in a tissue-specific manner.[22] Heterogeneous nuclear ribonucleoprotein (hnRNP) A1 regulates the Drosha processing of only miR-18a from the pri-miRNA that includes miR-17, -18a, -19a, -19b-1, -20a, and -92-1.[23] Cloning of miRNAs in human primary pulmonary artery smooth muscle cells stimulated with TGF-β demonstrated that mature miR-21 was increased.[24] Using qRT-PCR assays to quantify the pri-, pre- and mature miRNAs, these authors showed that TGF-β increased the mature and pre-miRNA in these cells without affecting levels of the pri-miRNA, effectively ruling out increased transcription as a possible explanation. The post-transcriptional regulation was due to a specific interaction between the pri-miRNAs, SMAD proteins, p68, and the Drosha complex.[24] Direct evidence for regulation of miRNA biogenesis at the nuclear-cytoplasmic transport level or Dicer processing steps is lacking However, indirect evidence has shown that levels of mature miRNAs often correlate with Dicer messenger RNA or protein expression[25,26] and that certain pre-miRNAs are localized to the nucleolus in cancer cell lines rather than being transported to the cytoplasm.[27]

Studying the regulation of the various steps in miRNA biogenesis may be achieved using northern blotting as these blots often show the presence or absence of the pri-, pre- and mature miRNA.[28,29] However, northern blots are cumbersome to perform and have poor sensitivity. It is also possible to include probes to the miRNA precursor on a cDNA miRNA array,[30] but these probes cannot distinguish the pri-miRNA from the pre-miRNA. As an alternative to northern blots or cDNA arrays, we propose using real-time qRT-PCR to quantify the pri-, pre- and mature miRNA.[31,32] Using this approach, one is able to quantify the relative expression or copy number of mature and pre-miRNAs from as little as 100 ng of total RNA and distinguish individual miRNAs among families of miRNA isoforms. This chapter will discuss the technology of qRT-PCR quantification of mature and pre-miRNAs and provide examples that use this technology.

7.2 Description of the Technology

MiRNAs are challenging molecules to amplify and quantify by RT-PCR since the mature miRNA is roughly the size of a standard PCR primer, and the miRNA precursors exist as a stable hairpin. However, despite these challenges, successful qRT-PCR applications have been developed to quantify both the mature[31,33] and precursor[32] miRNAs. We developed the first qRT-PCR assay for miRNA in 2004.[32] Since we were unsuccessful at developing technology to amplify the mature miRNA, we focused on an assay to amplify and quantify the miRNA precursors. Using RT and PCR primers that hybridize to the stem of the hairpin, a thermostable RT and a slightly elevated RT temperature, we were able to RT and amplify the miRNA precursors. Since the hairpin is contained within both the pri- and pre-miRNA, this assay in effect quantifies the pri-/pre-miRNA or miRNA precursors (Figure 7.1). Our thought at the time was that the levels of the miRNA precursors would directly correlate to the amount of active, mature miRNA. Using northern blots, we and others showed that the levels of pri-/pre-miRNA from qRT-PCR correlated with the levels of mature miRNA from northern blots.[32,34–37] However, as we began to profile the pri-, pre- and mature miRNAs by qRT-PCR in more and more situations, we found that in many cases the levels of the miRNA precursors did not correlate with the active, mature miRNA.[27]

As stated above, the premise behind the miRNA precursor assay is to use an RT primer (antisense PCR primer) to prime both the pri- and pre-miRNA (Figure 7.1). Using a pair of PCR primers that hybridize to the stem of the hairpin, the PCR will amplify both the pri- and pre-miRNA. Using a pair of PCR primers in which the sense primer hybridizes to a sequence upstream of the hairpin and the antisense primer, one is able to amplify the pri-miRNA alone (Figure 7.1). Subtracting the C_T of the pri-miRNA from the C_T of the pri-/pre-miRNA will provide the C_T for the pre-miRNA[35]:

$$C_{T\text{pri-/pre-}} - C_{T\text{pri-}} = C_{T\text{pre-}} \qquad \text{(Eq. 7.1)}$$

Figure 7.2 shows a good example of the ability of this approach to discriminate between the pri- and the pri-/pre-miRNAs. The real-time PCR plot shows a one C_T difference between the pri- and the pri-/pre-miRNAs (Figure 7.2a). A one C_T difference is exactly what one would expect if all of the pri-miRNA is processed to pre-miRNA, as in the case of miR-18 in HeLa and other cell lines (Figure 7.2). By designing TaqMan probes to hybridize to the precursor loop, one may distinguish individual miRNAs from a mixture

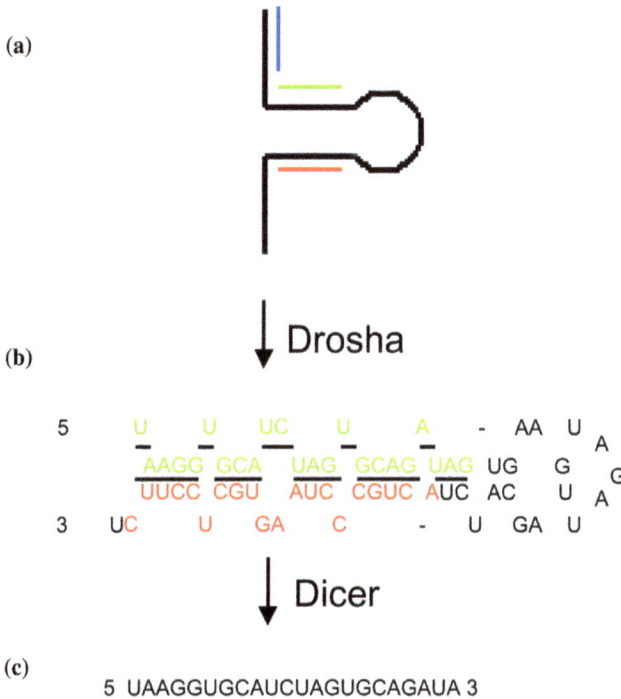

Figure 7.1. MiRNA biogenesis and placement of primers for qRT-PCR of miRNA precursors. MiRNA biogenesis is depicted in which the pri-miRNA (a) is processed by Drosha to the pre-miRNA (b), which is further processed by Dicer to the active, mature miRNA (c). The RT is primed with the antisense PCR primer (red). PCR of the pri-/pre-miRNA occurs using the sense (green) and antisense primers. PCR of the pri-miRNA occurs using the pri-miRNA specific primer (blue) and the antisense primer. Levels of the pre-miRNA are determined by subtracting the C_T of the pri-miRNA from the C_T of the pri-/pre-miRNA. Quantification of the mature miRNA is performed using the method described in the text.

of miRNA isoforms such as the let-7 family.[35] We have published TaqMan probes to the let-7 family and PCR primers to over 200 miRNA precursors.[35]

To assay the mature miRNA, we use the method developed by Applied Biosystems (Foster City, CA).[31] This assay utilizes a looped primer as the RT primer.[31] There are several advantages to the looped primer. First, by annealing a short RT priming sequence to the 3′ portion of the miRNA, it has better specificity for discriminating similar miRNAs. Secondly, its double-stranded stem structure inhibits hybridization of the RT primer to

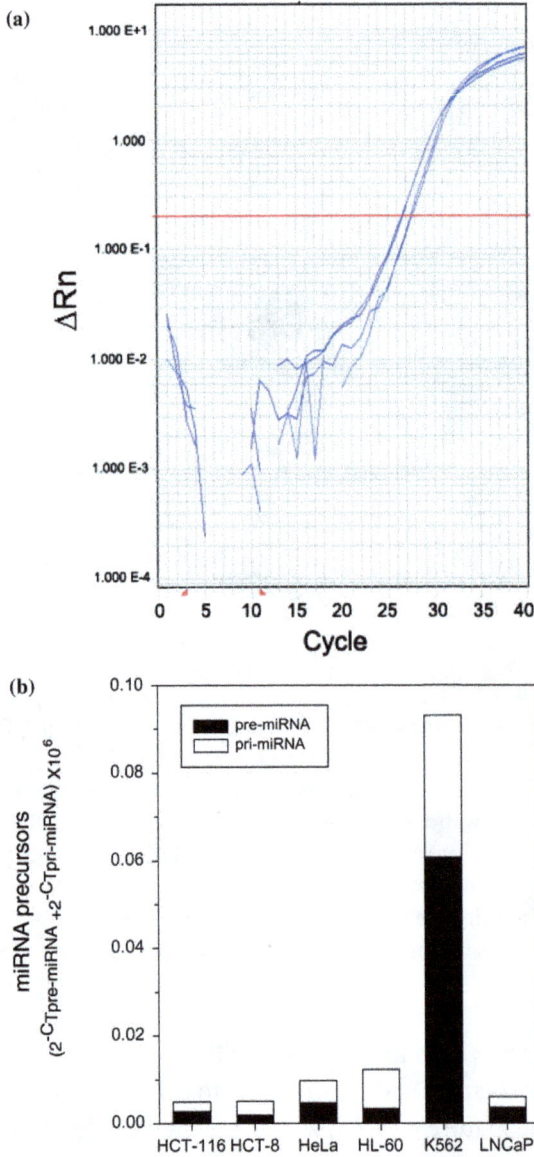

Figure 7.2.   Pri- and pri-/pre-miRNA expression in human cancer cell lines. (a) Total RNA from HeLa cells was converted to cDNA using gene specific primers as described in the text. The cDNA was amplified by real-time PCR using primers that anneal to the hairpin present in both the pri-miR-18 and pre-miR-18 (C_T = 26.6) or to the pri-miRNA only (C_T = 27.6). (b) Total miR-18 precursor expression (pri-miRNA + pre-miRNA combined) and individual expression (pri-miRNA or pre-miRNA) in six cancer cell lines.[32]

miRNA precursors and other long RNAs. Thirdly, the base stacking of the stem enhances the stability of miRNA and DNA hetero-duplexes, improving the RT efficiency for relatively short RT primers. Finally, the stem-loop structure, when unfolded, adds sequence downstream of the miRNA after RT. Following the RT, a forward primer containing a 5′ flap region and an antisense primer that hybridizes to the sequence contained within the looped primer are used for the PCR. The qRT-PCR assay to the mature miRNA uses TaqMan probes to distinguish individual miRNAs from a family of miRNA isoforms, has a 7-log dynamic range, and has a lower limit of sensitivity of 10 copies of mature miRNA or less.[31] By applying a standard curve, it is possible to present the gene expression data for both the pre- and mature miRNA as copy number per cell.[38]

7.3 Examples That Use the Technology

To demonstrate our ability to apply qRT-PCR to determine at which step in the miRNA biogenesis is being regulated, we performed an experiment similar to that of Kasashima *et al.*[39] This is to use 12-O-tetradecanoylphorbol-3-acetate (TPA) to induce the differentiation of the promyelocytic HL60 cells to monocyte/macrophage-like cells. HL60 cells were exposed to 16 nM TPA for 72 h. Differentiation was verified by cell morphology (not shown); undifferentiated HL60 cells grow in suspension while the differentiated cells become adherent. The expression of over 200 mature and pri-/pre-miRNAs was quantified using the assays described above.[38] TPA treatment increased the expression of the pri-/pre- and mature miR-21 by 5.2- and 6.5-fold, respectively. To determine if the increase in gene expression occurred at the transcriptional or post-transcriptional levels, we then measured the pri-miR-21 and determined the pre-miR-21 levels as described in the text. TPA increased the levels of the pri-, pre- and mature miR-21 by 4.4-, 7.3-, and 6.5-fold, respectively (Figure 7.3). Since TPA increased the levels of the pri-, pre- and mature miR-21, we conclude that the increase in miR-21 expression by TPA is due to enhanced transcription of the pri-miR-21.

7.4 Conclusions and Prospects

The intent of this chapter was to educate the reader on applying qRT-PCR as a means to determine which step in the miRNA biogenesis is being regulated.

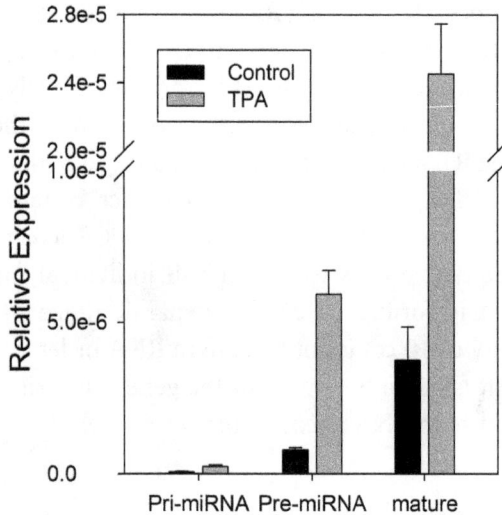

Figure 7.3.   **Change in miR-21 expression following TPA-induced differentiation of HL60 cells.** Undifferentiated promyelocytic HL60 cells were differentiated to myelocytic cells following exposure to 16 nM of TPA for 72 h. Total RNA was isolated from the cells and the primary precursor, precursor and mature miR-21 was assayed by the real-time PCR assays described in the text.

Real-time qRT-PCR has become the gold standard of gene expression quantification due to the sensitivity and specificity of the PCR. Various improvements in the technology have allowed one to successfully amplify and quantify the pri-, pre- and mature miRNA. By specifically assaying each of these molecules, it is possible to determine precisely at which step in the miRNA biogenesis is being regulated. An example of a miRNA whose expression is regulated post-transcriptionally was provided.

MiRNAs are clearly an important topic and this subject will continue to impact many biological studies in the near future. Understanding what regulates miRNA biogenesis is fundamental to our understanding of miRNA action. For example, if one were to identify a certain step in the miRNA biogenesis that is altered in a disease, it may be possible to develop drugs to correct this defect. While many studies have discovered defects in miRNA biogenesis without expression profiling,[18,19] we propose that expression profiling of both precursor and mature miRNAs using qRT-PCR will accelerate the discovery of situations in which miRNA biogenesis is being regulated. Expression profiling of both pre- and mature miRNA using qRT-PCR assays configured in a 384-well plate has been successfully used.[38] Unlike whole genome expression profiling that must be performed on a microchip, the current size of the human miRnome allows complete profiling in two 384-well plates. Using primers specific to the pri-, pre- and mature miRNA, it is possible to generate a large amount of data on individual miRNA precursors or mature miRNA for many miRNA genes. We believe that collection of such data may only be achieved using qRT-PCR.

Acknowledgments

Research supported by grant CA114304 (T.D.S.).

References

1. Lee R.C., Feinbaum R.L. and Ambros V. (1993). Tḫe *C. elegans* heterochronic gene lin-4 encodes small RNAs with antisense complementarity to lin-14. *Cell* **75**, 843–854.
2. Wightman B., Ha I. and Ruvkun G. (1993). Posttranscriptional regulation of the heterochronic gene lin-14 by lin-4 mediates temporal pattern formation in *C. elegans*. *Cell* **75**, 855–862.
3. Lagos-Quintana M., Rauhut R., Lendeckel W. *et al.* (2001). Identification of novel genes coding for small expressed RNAs. *Science* **294**, 853–858.
4. Lee R.C. and Ambros V. (2001). An extensive class of small RNAs in *Caenorhabditis elegans*. *Science* **294**, 862–864.
5. Lau N.C., Lim L.P., Weinstein E.G. *et al.* (2001). An abundant class of tiny RNAs with probable regulatory roles in *C. elegans*. *Science* **294**, 858–862.
6. Griffiths-Jones S. (2004). The microRNA registry. *Nucleic Acids Res* **32**, D109–111.
7. Lewis B.P., Burge C.B. and Bartel D.P. (2005). Conserved seed pairing, often flanked by adenosines, indicates that thousands of human genes are microRNA targets. *Cell* **120**, 15–20.
8. Kim Y.K. and Kim V.N. (2007). Processing of intronic microRNAs. *EMBO J* **26**, 775–783.
9. Lee Y., Kim M., Han J. *et al.* (2004). MicroRNA genes are transcribed by RNA polymerase II. *EMBO J* **23**, 4051–4060.
10. Lee Y., Ahn C., Han J. *et al.* (2003). The nuclear RNase III Drosha initiates microRNA processing. *Nature* **425**, 415–419.

11. Lund E., Guttinger S., Calado A. *et al.* (2004). Nuclear export of microRNA precursors. *Science* **303**, 95–98.

12. He L., Thomson J.M., Hemann M.T. *et al.* (2005). A microRNA polycistron as a potential human oncogene. *Nature* **435**, 828–833.

13. Calin G.A., Dumitru C.D., Shimizu M. *et al.* (2002). Frequent deletions and downregulation of microRNA genes miR15 and miR16 at 13q14 in chronic lymphocytic leukemia. *Proc Natl Acad Sci USA* **99**, 15524–15529.

14. Kim S., Lee U.J., Kim M.N. *et al.* (2008). MicroRNA miR-199a* regulates the MET proto-oncogene and the downstream extracellular signal-regulated kinase 2 (ERK2). *J Biol Chem* **283**, 18158–18166.

15. Lujambio A. and Esteller M. (2007). CpG island hypermethylation of tumor suppressor microRNAs in human cancer. *Cell Cycle* **6**, 1455–1459.

16. Lugli G., Torvik V.I., Larson J. *et al.* (2008). Expression of microRNAs and their precursors in synaptic fractions of adult mouse forebrain. *J Neurochem* **106**, 650–661.

17. Heo I., Joo C., Cho J. *et al.* (2008). Lin-28 mediates the terminal uridylation of let-7 precursor MicroRNA. *Mol Cell* **32**, 276–284.

18. Duan R., Pak C. and Jin P. (2007). Single nucleotide polymorphism associated with mature miR-125a alters the processing of pri-miRNA. *Hum Mol Genet* **16**, 1124–1131.

19. Jazdzewski K., Murray E.L., Franssila K. *et al.* (2008). Common SNP in pre-miR-146a decreases mature miR expression and predisposes to papillary thyroid carcinoma. *Proc Natl Acad Sci USA* **105**, 7269–7274.

20. Newman M.A., Thomson J.M. and Hammond S.M. (2008). Lin-28 interaction with the Let-7 precursor loop mediates regulated microRNA processing. *RNA* **8**, 1539–1549.

21. Viswanathan S.R., Daley G.Q. and Gregory R.I. (2008). Selective blockade of microRNA processing by Lin-28. *Science* **320**, 97–100.

22. Obernosterer G., Leuschner P.J., Alenius M. *et al.* (2006). Post-transcriptional regulation of microRNA expression. *RNA* **12**, 1161–1167.

23. Guil S. and Caceres J.F. (2007). The multifunctional RNA-binding protein hnRNP A1 is required for processing of miR-18a. *Nat Struct Mol Biol* **14**, 591–596.

24. Davis B.N., Hilyard A.C., Lagna G. *et al.* (2008). SMAD proteins control DROSHA-mediated microRNA maturation. *Nature* **454**, 56–61.

25. Chiosea S., Jelezcova E., Chandran U. *et al.* (2006). Up-regulation of dicer, a component of the microRNA machinery, in prostate adenocarcinoma. *Am J Pathol* **169**, 1812–1820.

26. Karube Y., Tanaka H., Osada H. *et al.* (2005). Reduced expression of Dicer associated with poor prognosis in lung cancer patients. *Cancer Sci* **96**, 111–115.
27. Lee E.J., Baek M., Gusev Y. *et al.* (2007). Systematic evaluation of microRNA processing patterns in tissues, cell lines, and tumors. *RNA* **14**, 35–42.
28. Ambros V., Lee R.C., Lavanway A. *et al.* (2003). MicroRNAs and other tiny endogenous RNAs in *C. elegans. Curr Biol* **13**, 807–818.
29. Thomson J.M., Newman M., Parker J.S. *et al.* (2006). Extensive post-transcriptional regulation of microRNAs and its implications for cancer. *Genes Dev* **20**, 2202–2207.
30. Liu C.G., Calin G.A., Meloon B. *et al.* (2004). An oligonucleotide microchip for genome-wide microRNA profiling in human and mouse tissues. *Proc Natl Acad Sci USA* **101**, 9740–9744.
31. Chen C., Ridzon D.A., Broomer A.J. *et al.* (2005). Real-time quantification of microRNAs by stem-loop RT-PCR. *Nucleic Acids Res* **33**, e179.
32. Schmittgen T.D., Jiang J., Liu Q. *et al.* (2004). A high-throughput method to monitor the expression of microRNA precursors. *Nucleic Acids Res* **32**, e43.
33. Shi R. and Chiang V.L. (2005). Facile means for quantifying microRNA expression by real-time PCR. *Biotechniques* **39**, 519–525.
34. Calin G.A., Liu C.G., Sevignani C. *et al.* (2004). MicroRNA profiling reveals distinct signatures in B cell chronic lymphocytic leukemias. *Proc Natl Acad Sci USA* **101**, 11755–11760.
35. Jiang J., Lee E.J., Gusev Y. *et al.* (2005). Real-time expression profiling of microRNA precursors in human cancer cell lines. *Nucleic Acids Res* **33**, 5394–5403.
36. Lee E.J., Gusev Y., Jiang J. *et al.* (2007). Expression profiling identifies microRNA signature in pancreatic cancer. *Int J Cancer* **120**, 1046–1054.
37. Yanaihara N., Caplen N., Bowman E. *et al.* (2006). Unique microRNA molecular profiles in lung cancer diagnosis and prognosis. *Cancer Cell* **9**, 189–198.
38. Schmittgen T.D., Lee E.J., Jiang J. *et al.* (2008). Real-time PCR quantification of precursor and mature microRNA. *Methods* **44**, 31–38.
39. Kasashima K., Nakamura Y. and Kozu T. (2004). Altered expression profiles of microRNAs during TPA-induced differentiation of HL-60 cells. *Biochem Biophys Res Commun* **322**, 403–410.

8

Single Nucleotide Polymorphisms in MicroRNAs and MicroRNA Binding Sites with Roles in Cancer

Lena J. Chin* and Frank J. Slack*,†*

Although they were first discovered in *C. elegans* and their exact mechanism of action is still not understood, microRNAs (miRNAs) have captured the imagination of many scientists because of their involvement in human cancers. MiRNAs regulate important cancer genes by binding to sites in their 3′ UTRs, and many miRNAs are located in fragile regions of the human genome, segments of chromosomes that are often involved in deletions, amplifications and breakpoints associated with cancer. In addition, there is growing evidence that mis-expression or loss of miRNAs is sufficient to cause cancer. There is also increasing evidence that DNA variations in miRNA genes and in miRNA binding sites are involved in cancer susceptibility.

8.1 MiRNAs as Oncogenes and Tumour Suppressors

MicroRNAs (miRNAs) have been shown to act as both oncogenes and tumour suppressors,[1] and we highlight some examples below. One of the

*Department of Molecular, Cellular, and Developmental Biology, Yale University, P.O. Box 208103, 266 Whitney Ave, New Haven, Connecticut, USA. †E-mail: frank.slack@yale.edu.

most studied examples of a miRNA tumour suppressor is *let-7*. *Let-7* genes are found in fragile sites associated with lung, breast, urothelial, cervical and ovarian cancers.[2] *Let-7* expression is also reduced in cancers, such as lung, breast and ovarian cancers.[3–6] Furthermore, *let-7* represses the expression of several known oncogenes, such as *KRAS, Myc* and *HMGA2*.[5,7,8] *MiR-15* and *miR-16*, another well-studied family of tumour suppressor miRNAs, are often deleted or repressed in chronic lymphocytic leukaemia (CLL) and prostate and pituitary cancers.[9–11] MiR-15 and miR-16 regulate *CCND1, WNT3A* and *BCL2* expression.[9,12–15] *CCND3, CCNE1* and *CDK6* have also been found to be targets of the miR-16 family of miRNAs.[14]

There are also several commonly studied miRNA oncogenes. For example, *miR-155* has been found to be overexpressed in numerous different types of cancers, including B-cell lymphoma, Hodgkin lymphoma, and breast, lung and colon cancers.[4,16–20] MiR-155 is known to target the *TP53INP1, AGTR1* and *AID* genes.[21–23] The miR-17-92-1 cluster of miRNAs is another example of miRNAs that suppress oncogenes and this cluster is overexpressed in B-cell lymphoma and lung, breast, colon, pancreatic and prostate cancers.[19,24,25] This cluster consists of miR-17-5p, miR-17-3p, miR-18, miR-19a, miR-20a, miR-19b and miR-92-1. MiR-17-5p and miR-20a have been found to repress *E2F1*,[26] *BRCA1* and TGFBR2.[19,27]

8.2 MiRNA Profiling

With the development of technologies that aim to look at the expression levels of hundreds of miRNAs at the same time, a number of groups looked at miRNA profiles in different cancers. Calin *et al.* were the first to show that their miRNA microarray expression profiles could differentiate between B cell chronic lymphocyte leukaemia cells and normal cells.[28] Furthermore, they classified CLL samples into two different groups based on their miRNA profiles, and these profiles corresponded to high or low levels of a protein that is associated with a positive prognosis at low levels.[28] Johnson *et al.*[29] and Takamazawa *et al.*[6] used miRNA microarrays to demonstrate that *let-7* levels were low in non-small cell lung cancers. Shortly thereafter, Lu *et al.* developed a method for bead-based miRNA profiling that they proposed to be more sensitive since the hybridizations were taking place in solution.[30] Employing this technique on 20 different cancers, they found that each cancer had a specific miRNA profile and that most poorly differentiated tumours could be classified to their tissues of origin based on their miRNA expression levels.[30] Similarly, in a study of the NCI-60 panel of 59 cell lines, there was a significant reduction in overall miRNA expression in the cancer cell lines as compared to normal tissues of corresponding tissue origin.[31]

On the other hand, analysis of miRNA expression in 363 solid tumours from breast, colon, lung, pancreas, prostate and stomach cancers did not show such extensive miRNA downregulation in tumours.[19] Furthermore, when a miRNA was overexpressed, it was almost always the mature miRNA that was overexpressed as compared to the pre-miRNA. Volinia *et al.*, however, did find differential miRNA expression for each tissue type.[19] Through various techniques, numerous groups have now shown that different cancer types have distinct miRNA profiles; the list of publications covering this area is too extensive to list here. Furthermore, new methods are continuously being developed for determining miRNA expression profiles. As techniques continue to improve, it is likely that a miRNA profile will be identified for every cancer type. These cancer-specific miRNA signatures will likely be useful for classifying tumour origin of poorly differentiated tumours, and being able to determine the tissue of origin of these tumours will be valuable in determining a patient's course of treatment.

8.3 Dual Roles for MiRNAs as Oncogenes and Tumour Suppressors

KRAS is a very well-studied oncogene. However, several studies in mice and of lung cancer tumour samples identified roles for *KRAS* as a tumour suppressor gene.[32–35] Interestingly, just as some protein-coding genes may not be only tumour suppressor genes or oncogenes, miRNAs may not either. For example, *mir-16* is thought to be a tumour suppressor gene and regulates oncogenes. However, in identifying a miRNA signature of serous ovarian cancer, Nam *et al.* found that miR-16 was more highly expressed in the majority of tumour samples than in the corresponding normal ovarian tissue samples.[36] Similarly, in four cervical cancer samples, Wang *et al.* determined that miR-16 was highly expressed as compared to paired normal cervical tissue samples.[37] *MiR-17-5p* and *miR-20a*, on the other hand, are commonly thought to function as oncogenes, but there is evidence that these miRNAs may also be tumour suppressors in some cellular contexts. Yu *et al.* found that *miR-17-5p* and *miR-20a* were expressed at low levels in the majority of breast cancer cell lines examined as compared to controls. Furthermore, they determined that these miRNAs repressed *cyclin D1* through a conserved binding site in the *cyclin D1* 3' UTR.[38]

8.4 Single Nucleotide Polymorphisms (SNPs) Associated with MiRNA Genes

With the realisation of the importance of miRNAs in the regulation of gene expression, researchers began to look at the role of naturally occurring variations associated with miRNAs within the human genome. These include variation in both the miRNAs themselves and in their predicted target sites.

While some single nucleotide polymorphisms (SNPs) may not have an effect on miRNA expression or gene regulation, there have been several studies that identified polymorphisms that do have an effect on the miRNA (Figure 8.1). The first miRNA-associated SNP was identified in two chronic lymphocytic leukaemia patients. Calin *et al.* identified a germline C-to-T transition in pri-*mir-16-1*. These patients had low levels of miR-16-1, and cells transfected with a reporter construct with the variant allele expressed at significantly lower levels of miR-16-1 as compared to a reporter construct with the wild-type allele.[39] Furthermore, in sequencing 17 miRNA genes associated with regulating breast cancer genes in breast cancer patients, rare SNPs in pre-*miR-30c-1* and pri-*miR-17* were identified. These SNPs led to changes in secondary structure of the pre-miRNA and altered miRNA expression.[27] Iwai *et al.* similarly discovered that a SNP in the mature miR-30c-2 alters the base pairing in the stem of the hairpin structure. This polymorphism is predicted to alter miR-30c-2 processing and target mRNA selection because an expression vector containing miR-30c-2 with the variant allele gave rise to more improperly processed RNA products than miR-30c-2 with the wild-type allele.[40] Likewise, a SNP at the 8th nucleotide of the mature miR-125a affected pri-*miR-125a*, a *C. elegans lin-4* homologue, from being processed into pre-*miR-125a*. When the variant allele was present with a compensatory mutation to maintain the normal secondary structure, the proper miRNA processing was restored.[41]

miRNA-associated SNPs

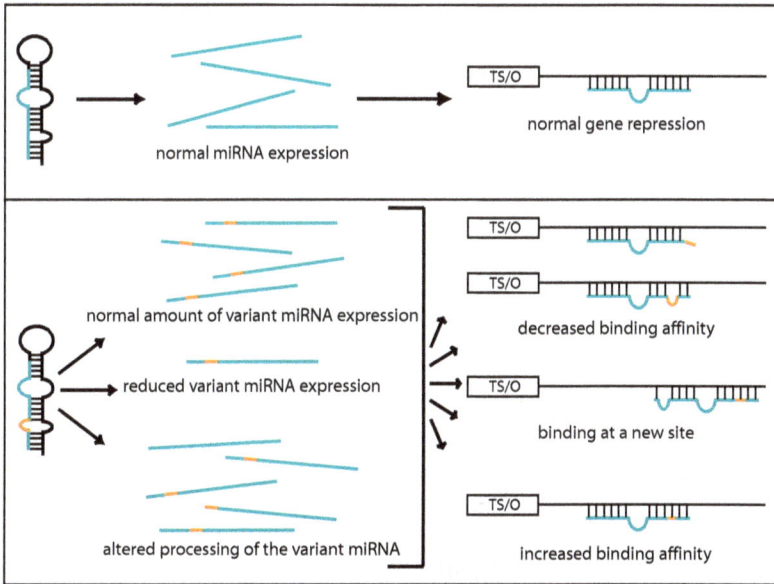

Figure 8.1. MiRNA-associated SNPs can alter target gene expression. A SNP (orange) in the mature miRNA region (blue) of a pre-miRNA is depicted here. A variant SNP allele in a miRNA gene may change the sequence of the miRNA produced, affect the amount of mature miRNA expressed, and alter the processing of the miRNA and give rise to products of different sizes or sequences. The altered miRNAs may have a different binding affinity as compared to the unaltered miRNA to the target tumour suppressor gene (TS) or oncogene (O) 3′ UTR, or may bind to new target sites in the 3′ UTR.

Another example of a SNP altering miRNA levels was found by Jazdzewski *et al.* They detected a known SNP on the passenger strand of pre-*miR-146a* in 15 papillary thyroid carcinoma samples. Through the use of plasmids expressing pri-*miR-146a* with either the G or C allele, it was shown that the C allele led to a reduction in mature miR-146a levels, but this was not the result of the C allele interfering with Dicer processing.[42] Heterozygosity at this SNP in pre-*miR-146a* was found to be associated with an increased risk of developing papillary thyroid carcinoma (odds ratio (OR) = 1.62). Interestingly, in this case, the homozygous genotypes appeared to play a protective role.[42]

Lastly, another study found other SNPs associated with non-small cell lung cancer (NSCLC). In this study of Chinese NSCLC patients, Hu *et al.*

identified two SNPs that are associated with patient survival. Regarding the *miR-196a-2* SNP, patients homozygous for the C allele had a statistically significant reduction in survival.[43] This SNP in *mir-196a-2* also appeared to adversely affect miR-196a processing. The presence of the variant allele at this SNP reduced production of miR-196a from pre-*miR-196a*.[43]

Saunders *et al.* determined that there were ~1.3 SNPs/kilobase (kb) of pre-miRNA; however, only 3 of the 65 SNPs were in the seed region of the miRNA. Furthermore, only ~10% of human pre-miRNAs had known SNPs.[44]

Therefore, there are clearly variants in miRNAs that disrupt the normal processing of pri- and pre-miRNAs, whether through complete inhibition or altered miRNA processing. These SNPs may also result in the improper binding of the miRNA to its target mRNA binding site or the miRNA binding to a new site, which could change the overall gene expression patterns (Figure 8.1).

8.5 SNPs in MiRNA Binding Sites

The identification of SNPs in miRNA binding sites has become a new area of study that looks for markers of cancer risk[45] and which tries to gain insight into miRNA-mediated gene regulation.[22,46–48] SNPs in 3′ UTRs may create new miRNA binding sites, weaken or disrupt miRNA binding, or improve miRNA binding (Figure 8.2). The first miRNA target site-associated variant was identified in two Tourette's syndrome patients. The SNP is in a predicted miR-189 binding site in the *SLITRK1* 3′ UTR, and the presence of the variant allele in a luciferase reporter resulted in slight, but significant, increase in luciferase repression as compared to the reference allele.[46] Thus, the polymorphism improved miR-189-mediated repression.

Furthermore, research has shown that it is not only SNPs within miRNA binding sites that may be important but that SNPs near miRNA binding sites could also be influential. Mishra *et al.* identified a SNP 14 nucleotides downstream of a putative miR-24 binding site in the *dihydrofolate reductase* (*DHFR*) gene. The presence of the variant allele of this SNP resulted in the loss of reporter repression and an increase in *DHFR* mRNA stability.[48] Thus, SNPs in 3′ UTRs have the potential to alter gene expression simply through disrupting one miRNA target site or creating a new miRNA target site (Figure 8.2).

miRNA binding site-associated SNPs

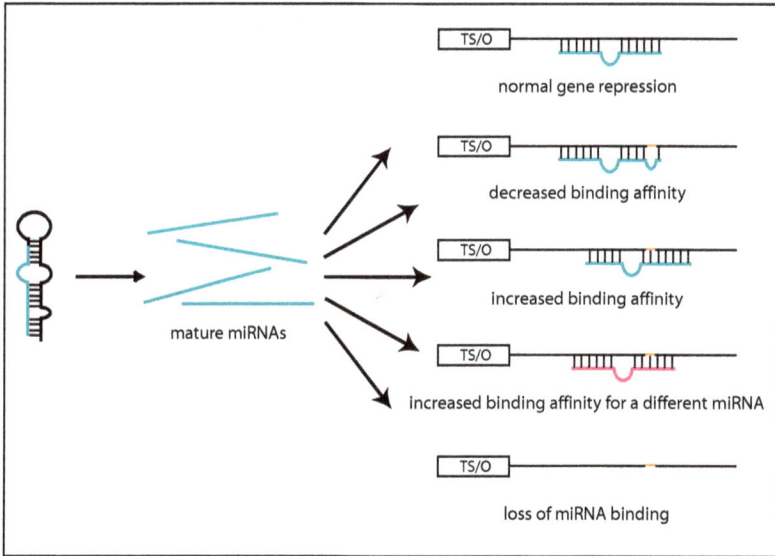

Figure 8.2. MiRNA binding site-associated SNPs can alter miRNA-mediated target tumour suppressor gene (TS) or oncogene (O) expression. The mature miRNA (blue) is excised from the hairpin pre-miRNA. The presence of a variant SNP allele in the miRNA's target gene may alter the miRNA's binding affinity as compared to the unaltered target site, increase the binding affinity for a different miRNA (pink), or completely disrupt miRNA binding at that site.

Interestingly, several studies have found that the overall frequency of single nucleotide polymorphisms occurring in putative miRNA target sites is lower than in other parts of the 3′ UTR. Chen and Rajewsky found that there were 0.5 SNPs/kb of conserved miRNA target sites, as compared to 0.73 SNPs/kb of conserved control sites.[49] Saunders *et al.*, however, identified ~1.9 SNPs/kb in computationally-predicted, conserved target sites and ~2.7 SNPs/kb in flanking regions.[44] The differences between these studies may be the result of several factors, including the version of the SNP database used, the types of SNPs included and the species and criteria used in determining conservation. Lastly, when comparing SNPs in the seed region of predicted miRNA target sites, the expected trend continues. SNPs occurred less frequently in the seed regions than in the whole 3′ UTRs.[49,50] Therefore, it appears that there is strong negative selection against SNPs in important regions. Some, in fact, believe that most of these miRNA-related SNPs are not playing an important role in miRNA expression[40] or miRNA-mediated target regulation.[44]

8.6 MiRNA-Associated SNPs Associated with Cancer

Polymorphisms were found in the miR-221/222 and miR-146a/b target sites in the *KIT* 3' UTR of papillary thyroid carcinoma tumours.[47] The alteration of the miR-221/222 binding site was predicted to cause conformational changes that increased free energy; therefore, the miRNAs could potentially no longer bind at this site or could bind to this site with lower affinity. The SNP in the miR-146a/b binding site was thought to result in the miRNA(s) binding to a different region of the 3' UTR.[47] Sethupathy *et al.* also found that a SNP in the miR-155 binding site in the *AGTR1* 3' UTR resulted in selective miR-155-mediated repression. A reporter with the variant C allele was not downregulated by miR-155, where as the wild-type 3' UTR was downregulated by miR-155.[22]

MiRNA-binding SNPs are beginning to be found to be associated with increased risk of numerous diseases, including human cancers. For instance, through a study of the 3' UTRs of 104 colorectal cancer-related genes, Landi *et al.* identified two SNPs in predicted miRNA binding sites of the *CD86 antigen* and *insulin receptor* genes associated with colorectal cancer risk (OR = 1.33 and OR = 2.93, respectively).[51] The reference/wild-type allele of a SNP in the *trinucleotide repeat-containing gene 6b* gene was linked to aggressive prostate cancer (OR = 1.18).[52] Furthermore, a study of Swedish breast cancer cases revealed a SNP in a putative miRNA binding site within the integrin gene *ITGB4* 3' UTR. Carriers of the variant allele had poorer prognoses and more aggressive tumours. The SNP, however, was not associated with breast cancer risk.[53]

Lung cancer is the leading cause of cancer deaths in both the United Kindgom and the United States.[54,55] There are two major classes of lung cancer: non-small cell lung cancer (NSCLC) and small cell lung cancer (SCLC). NSCLC makes up the majority of lung cancer cases. According to the American Cancer Society in 2008, lung cancer was expected to make up 15% of all new cancer diagnoses, but 29% of all cancer deaths. This discrepancy is because a lung cancer diagnosis usually comes with a very poor prognosis; the overall 5-year survival rate for all stages is 15%. However, early detection greatly improves the 5-year survival rate to 49% for patients with localised cancer. The problem is that only 16% of lung cancer cases are diagnosed at such an early stage.[56] Numerous studies have been published examining polymorphisms and the risk of lung cancer. Much of this work has been focused on DNA repair pathways.[57–70] In addition, some studies conflict with one another regarding the associations between the polymorphisms and lung cancer, such as studies involving a SNP in the promoter of *MDM2*;[71,72] these differences may be a result of the populations, the number of cases, or the distribution of types of NSCLC sampled. Overall, there still is not a good marker that is utilised clinically for identifying lung cancer risk. Developing tools to identify the disease earlier could have a tremendous impact on improving patient outcome. Furthermore, the identification of such mutations in any of these genes could also shed light onto cell cycle regulation, tumour development, and/or miRNA-mediated gene regulation. A previously identified SNP that affects miR-192a2 processing[43] has since been shown to increase the risk of developing NSCLC in Chinese patients, especially in smokers.[73] A SNP in a predicted *let-7* miRNA binding site in the *KRAS* 3′ UTR was identified with an increased risk of developing non-small cell lung cancer in moderate smokers (OR = 1.36 − 2.3)[45] and could be useful in determining cancer risk in this population.

Therefore, the identification of miRNA-associated SNPs that are associated with cancer risk will likely prove to be useful for determining people with a greater risk of developing particular types of cancer. Thus, if people can be more thoroughly monitored for tumour development, perhaps cancers that would normally be detected at a later stage will now be found earlier.

8.7 Conclusions

MiRNA-associated SNPs can play several roles in cancer. As previously mentioned, these SNPs can lead to the misregulation of gene expression; however, they can also be useful for identifying people at greater risk for a specific type of cancer, for determining prognostic outlooks, or even for identifying people who are less likely to develop a type of cancer. As this field continues to be studied, researchers are likely to continue to find miRNA-related SNPs linked with cancer. As more and more studies of the polymorphisms in our genome are finished, hopefully we will be able to better identify people at an increased risk of cancer through genotyping particular sets of SNPs or have a better understanding of the characteristics of the tumour present.

References

1. Esquela-Kerscher A. and Slack F. (2006). Oncomirs — MicroRNAs with a role in cancer. *Nat Rev Cancer* **6**, 259–269.
2. Calin G.A., Sevignani C., Dumitru C.D. *et al.* (2004). Human microRNA genes are frequently located at fragile sites and genomic regions involved in cancers. *Proc Natl Acad Sci USA* **101**, 2999–3004.
3. Iorio M.V., Visone R., Di Leva G. *et al.* (2007). MicroRNA signatures in human ovarian cancer. *Cancer Res* **67**, 8699–8707.
4. Iorio M.V., Ferracin M., Liu C.G. *et al.* (2005). MicroRNA gene expression deregulation in human breast cancer. *Cancer Res* **65**, 7065–7070.
5. Johnson S.M., Grosshans H., Shingara J. *et al.* (2005). RAS is regulated by the let-7 microRNA family. *Cell* **120**, 635–647.
6. Takamizawa J., Konishi H., Yanagisawa K. *et al.* (2004). Reduced expression of the let-7 microRNAs in human lung cancers in association with shortened postoperative survival. *Cancer Res* **64**, 3753–3756.
7. Kumar M.S., Lu J., Mercer K.L. *et al.* (2007). Impaired microRNA processing enhances cellular transformation and tumorigenesis. *Nat Genet* **39**, 673–677.
8. Mayr C., Hemann M.T. and Bartel D.P. (2007). Disrupting the pairing between let-7 and Hmga2 enhances oncogenic transformation. *Science* **315**, 1576–1579.
9. Bonci D., Coppola V., Musumeci M. *et al.* (2008). The miR-15a-miR-16-1 cluster controls prostate cancer by targeting multiple oncogenic activities. *Nat Med* **14**, 1271–1277.
10. Bottoni A., Piccin D., Tagliati F. *et al.* (2005). miR-15a and miR-16-1 down-regulation in pituitary adenomas. *J Cell Physiol* **204**, 280–285.
11. Calin G.A., Dumitru C.D., Shimizu M. *et al.* (2002). Frequent deletions and down-regulation of microRNA genes miR15 and miR16 at 13q14 in chronic lymphocytic leukemia. *Proc Natl Acad Sci USA* **99**, 15524–15529.
12. Chen R.W., Bemis L.T., Amato C.M. *et al.* (2008). Truncation in CCND1 mRNA alters miR-16-1 regulation in mantle cell lymphoma. *Blood* **112**, 822–829.

13. Cimmino A., Calin G.A., Fabbri M. *et al.* (2005). miR-15 and miR-16 induce apoptosis by targeting BCL2. *Proc Natl Acad Sci USA* **102**, 13944–13949.
14. Liu Q., Fu H., Sun F. *et al.* (2008). miR-16 family induces cell cycle arrest by regulating multiple cell cycle genes. *Nucleic Acids Res* **36**, 5391–5404.
15. Xia L., Zhang D., Du R. *et al.* (2008). miR-15b and miR-16 modulate multidrug resistance by targeting BCL2 in human gastric cancer cells. *Int J Cancer* **123**, 372–379.
16. Eis P.S., Tam W., Sun L. *et al.* (2005). Accumulation of miR-155 and BIC RNA in human B-cell lymphomas. *Proc Natl Acad Sci USA* **102**, 3627–3632.
17. Kluiver J., Poppema S., de Jong D. *et al.* BIC and miR-155 are highly expressed in Hodgkin, primary mediastinal and diffuse large B-cell lymphomas. *J Pathol* **207**, 243–249.
18. van den Berg A., Kroesen B.J., Kooistra K. *et al.* (2003). High expression of B-cell receptor inducible gene BIC in all subtypes of Hodgkin lymphoma. *Gene Chromosomes Canc* **37**, 20–28.
19. Volinia S., Calin G.A., Liu C.G. *et al.* (2006). A microRNA expression signature of human solid tumors defines cancer gene targets. *Proc Natl Acad Sci USA* **103**, 2257–2261.
20. Yanaihara N., Caplen N., Bowman E. *et al.* (2006). Unique microRNA molecular profiles in lung cancer diagnosis and prognosis. *Cancer Cell* **9**, 189–198.
21. Gironella M., Seux M., Xie M.J. *et al.* (2007). Tumor protein 53-induced nuclear protein 1 expression is repressed by miR-155, and its restoration inhibits pancreatic tumor development. *Proc Natl Acad Sci USA* **104**, 16170–16175.
22. Sethupathy P., Borel C., Gagnebin M. *et al.* (2007). Human microRNA-155 on chromosome 21 differentially interacts with its polymorphic target in the AGTR1 3′ untranslated region: A mechanism for functional single-nucleotide polymorphisms related to phenotypes. *Am J Hum Genet* **81**, 405–413.
23. Teng G., Hakimpour P., Landgraf P. *et al.* (2008). MicroRNA-155 is a negative regulator of activation-induced cytidine deaminase. *Immunity* **28**, 621–629.
24. Hayashita Y., Osada H., Tatematsu Y. *et al.* (2005). A polycistronic microRNA cluster, miR-17-92, is overexpressed in human lung cancers and enhances cell proliferation. *Cancer Res* **65**, 9628–9632.
25. He L., Thomson J.M., Hemann M.T. *et al.* (2005). A microRNA polycistron as a potential human oncogene. *Nature* **435**, 828–833.
26. O'Donnell K.A., Wentzel E., Zeller K.I. *et al.* (2005). c-Myc-regulated microRNAs modulate E2F1 expression. *Nature* **435**, 839–843.
27. Shen J., Ambrosone C.B. and Zhao H. (2009). Novel genetic variants in microRNA genes and familial breast cancer. *Int J Cancer* **124**, 1178–1182.

28. Calin G.A., Liu C.G., Sevignani C. *et al.* (2004). MicroRNA profiling reveals distinct signatures in B cell chronic lymphocytic leukemias. *Proc Natl Acad Sci USA* **101**, 11755–11760.

29. Johnson C.D., Esquela-Kerscher A., Stefani G. *et al.* (2007). The let-7 microRNA represses cell proliferation pathways in human cells. *Cancer Res* **67**, 7713–7722.

30. Lu J., Getz G., Miska E.A. *et al.* (2005). MicroRNA expression profiles classify human cancers. *Nature* **435**, 834–838.

31. Gaur A., Jewell D.A., Liang Y. *et al.* (2007). Characterization of microRNA expression levels and their biological correlates in human cancer cell lines. *Cancer Res* **67**, 2456–2468.

32. Hegi M.E., Devereux T.R., Dietrich W.F. *et al.* (1994). Allelotype analysis of mouse lung carcinomas reveals frequent allelic losses on chromosome 4 and an association between allelic imbalances on chromosome 6 and K-ras activation. *Cancer Res* **54**, 6257–6264.

33. James R.M., Arends M.J., Plowman S.J. *et al.* (2003). K-ras proto-oncogene exhibits tumor suppressor activity as its absence promotes tumorigenesis in murine teratomas. *Mol Cancer Res* **1**, 820–825.

34. Li J., Zhang Z., Dai Z. *et al.* (2003). LOH of chromosome 12p correlates with Kras2 mutation in non-small cell lung cancer. *Oncogene* **22**, 1243–1246.

35. Zhang Z., Wang Y.Q., Vikis H.G. *et al.* (20010. Wildtype Kras2 can inhibit lung carcinogenesis in mice. *Nat Genet* **29**, 25–33.

36. Nam E.J., Yoon H., Kim S.W. *et al.* (2008). MicroRNA expression profiles in serous ovarian carcinoma. *Clin Cancer Res* **14**, 2690–2695.

37. Wang X., Tang S., Le S. *et al.* (2008). Aberrant expression of oncogenic and tumor-suppressive microRNAs in cervical cancer is required for cancer cell growth. *PLoS ONE* **3**, e2557.

38. Yu Z., Wang C., Wang M. *et al.* (2008). A cyclin D1/microRNA 17/20 regulatory feedback loop in control of breast cancer cell proliferation. *J Cell Biol* **182**, 509–517.

39. Calin G.A., Ferracin M., Cimmino A. *et al.* (2005). A microRNA signature associated with prognosis and progression in chronic lymphocytic leukemia. *N Engl J Med* **353**, 1793–1801.

40. Iwai N. and Naraba H. (2005). Polymorphisms in human pre-miRNAs. *Biochem Biophys Res Commun* **331**, 1439–1444.

41. Duan R., Pak C. and Jin P. (2007). Single nucleotide polymorphism associated with mature miR-125a alters the processing of pri-miRNA. *Human Mol Gen* **16**, 1124–1131.

42. Jazdzewski K., Murray E.L., Franssila K. *et al.* (2008). Common SNP in pre-miR-146a decreases mature miR expression and predisposes to papillary thyroid carcinoma. *Proc Natl Acad Sci USA* **105**, 7269–7274.
43. Hu Z., Chen J., Tian T. *et al.* (2008). Genetic variants of miRNA sequences and non-small cell lung cancer survival. *J Clin Invest* **118**, 2600–2608.
44. Saunders M.A., Liang H. and Li W.H. (2007). Human polymorphism at microRNAs and microRNA target sites. *Proc Natl Acad Sci USA* **104**, 3300–3305.
45. Chin L.J., Ratner E., Leng S. *et al.* (2008). A SNP in a let-7 microRNA complementary site in the KRAS 3′ untranslated region increases non-small cell lung cancer risk. *Cancer Res* **68**, 8535–8540.
46. Abelson J.F., Kwan K.Y., O'Roak B.J. *et al.* (2005). Sequence variants in SLITRK1 are associated with Tourette's syndrome. *Science* **310**, 317–320.
47. He H., Jazdzewski K., Li W. *et al.* (2005). The role of microRNA genes in papillary thyroid carcinoma. *Proc Natl Acad Sci USA* **102**, 19075–19080.
48. Mishra P.J., Humeniuk R., Longo-Sorbello G.S. *et al.* (2007). A miR-24 microRNA binding-site polymorphism in dihydrofolate reductase gene leads to methotrexate resistance. *Proc Natl Acad Sci USA* **104**, 13513–13518.
49. Chen K. and Rajewsky N. (2006). Natural selection on human microRNA binding sites inferred from SNP data. *Nat Genet* **38**, 1452–1456.
50. Yu Z., Li Z., Jolicoeur N. *et al.* (2007). Aberrant allele frequencies of the SNPs located in microRNA target sites are potentially associated with human cancers. *Nucleic Acids Res* **35**, 4535–4541.
51. Landi D., Gemignani F., Naccarati A. *et al.* (2007). Polymorphisms within micro-RNA-binding sites and risk of sporadic colorectal cancer. *Carcinogenesis* **29**, 579–584.
52. Sun J., Zheng S.L., Wiklund F. *et al.* (2006). Sequence variants at 22q13 are associated with prostate cancer risk. *Cancer Res* **69**, 10–15.
53. Brendle A., Lei H., Brandt A. *et al.* (2008). Polymorphisms in predicted microRNA-binding sites in integrin genes and breast cancer: ITGB4 as prognostic marker. *Carcinogenesis* **29**, 1394–1399.
54. Cancer Research UK. (2007). News and Resources Centre CancerStats: Lung Cancer. 1–14.
55. Jemal A., Siegel R., Ward E. *et al.* (2008). Cancer statistics, 2008. *CA Cancer J Clin* **58**, 71–96.
56. American Cancer Society. (2008). *Cancer Facts and Figures 2008*. American Cancer Society, Atlanta.
57. Bai Y., Xu L., Yang X. *et al.* (2007). Sequence variations in DNA repair gene XPC is associated with lung cancer risk in a Chinese population: A case-control study. *BMC Cancer* **7**, 81.

58. Biros E., Kalina I., Kohút A. *et al.* (2001). Germ line polymorphisms of the tumor suppressor gene p53 and lung cancer. *Lung Cancer* **31**, 157–162.

59. Chang J.S., Wrensch M.R., Hansen H.M. *et al.* (2009). Base excision repair genes and risk of lung cancer among San Francisco Bay Area Latinos and African Americans. *Carcinogenesis* **30**, 78–87.

60. Hao B., Miao X., Li Y. *et al.* (2006). A novel T-77C polymorphism in DNA repair gene XRCC1 contributes to diminished promoter activity and increased risk of non-small cell lung cancer. *Oncogene* **25**, 3613–3620.

61. Hu Z., Liu H., Wang H. *et al.* (2008). Tagging single nucleotide polymorphisms in phosphoinositide-3-kinase-related protein kinase genes involved in DNA damage 'checkpoints' and lung cancer susceptibility. *Clin Cancer Res* **14**, 2887–91.

62. Hu Z., Shao M., Yuan J. *et al.* (2006). Polymorphisms in DNA damage binding protein 2 (DDB2) and susceptibility of primary lung cancer in the Chinese: A case-control study. *Carcinogenesis* **27**, 1475–1480.

63. Kanzaki H., Ouchida M., Hanafusa H. *et al.* (2008). The association between RAD18 Arg302Gln polymorphism and the risk of human non-small-cell lung cancer. *J Cancer Res Clin Oncol* **134**, 211–217.

64. Kiyohara C. and Yoshimasu K. (2007). Genetic polymorphisms in the nucleotide excision repair pathway and lung cancer risk: a meta-analysis. *Int J Med Sci* **4**, 59–71.

65. Miao R., Gu H., Liu H. *et al.* (2008). Tagging single nucleotide polymorphisms in MBD4 are associated with risk of lung cancer in a Chinese population. *Lung Cancer* **62**, 281–286.

66. Rudd M.F., Webb E.L., Matakidou A. *et al.* (2006). Variants in the GH-IGF axis confer susceptibility to lung cancer. *Genome Res* **16**, 693–701.

67. Sakiyama T., Kohno T., Mimaki S. *et al.* (2005). Association of amino acid substitution polymorphisms in DNA repair genes TP53, POLI, REV1 and LIG4 with lung cancer risk. *Int J Cancer* **114**, 730–737.

68. Wang L., Lu J., An J. *et al.* (2007). Polymorphisms of cytosolic serine hydroxymethyltransferase and risk of lung cancer: A case-control analysis. *Lung Cancer* **57**, 143–151.

69. Zhou W., Liu G., Park S. *et al.* (2005) Gene-smoking interaction associations for the ERCC1 polymorphisms in the risk of lung cancer. *Cancer Epidemiol Biomarkers Prev* **14**, 491–496.

70. Zhou W., Liu G., Miller D.P. *et al.* (2002). Gene-environment interaction for the ERCC2 polymorphisms and cumulative cigarette smoking exposure in lung cancer. *Cancer Res* **62**, 1377–1381.

71. Lind H., Zienolddiny S., Ekstrøm P.O. *et al.* (2006). Association of a functional polymorphism in the promoter of the MDM2 gene with risk of nonsmall cell lung cancer. *Int J Cancer* **119**, 718–721.

72. Pine S.R., Mechanic L.E., Bowman E.D. *et al.* (2002). MDM2 SNP309 and SNP354 are not associated with lung cancer risk. *Cancer Epidemiol Biomarkers Prev* **15**, 1559–1561.

73. Tian T., Shu Y., Chen J. *et al.* (2009). A functional genetic variant in microRNA-196a2 is associated with increased susceptibility of lung cancer in Chinese. *Cancer Epidemiol Biomarkers Prev* **18**, 1183–1187.

9

MicroRNAs as Potential Diagnostics and Therapeutics

Trupti Paranjape,[*,†] *Jae Choi*[†] *and Joanne B. Weidhaas*[*,†,‡]

Although major advances have been achieved in our understanding of cancer biology as well as in the development of new-targeted therapies, the progress in developing improved early diagnosis and screening tests has been inadequate. This results in most cancers being diagnosed in advanced stages leading to poor outcomes. There is intense research seeking specific molecular changes that are able to identify patients with early cancer or precursor lesions. Recently, a novel class of global gene regulators called microRNAs (miRNAs) was identified as important in cancer. MiRNAs that are overexpressed in cancer function as oncogenes, and miRNAs with tumour suppressor activity in normal tissue may be downregulated in cancer. MiRNA expression data in various cancers demonstrate that each tissue has its unique miRNA expression patterns and that cancer cells have different miRNA profiles compared with normal cells, thus underscoring the tremendous diagnostic potential of miRNAs in cancer. Furthermore, because miRNAs regulate hundreds of gene targets critical in cell growth and survival they are also potentially powerful new therapeutics. These

* Department of Therapeutic Radiology, †Yale University School of Medicine, New Haven, CT, USA. ‡E-mail: joanne.weidhaas@yale.edu.

unique properties of miRNAs make them some of the most promising advances towards personalised care for individual patients in the future.

9.1 MicroRNAs in Cancer Diagnostics

In the United States cancer is the second leading cause of death accounting for nearly one-quarter of deaths, exceeded only by heart diseases. In the United Kingdom cancer stands to be the third leading cause of death next to heart disease and stroke. Worldwide cancer is the third leading cause of mortality after cardiovascular and infectious diseases. It was estimated that in 2008 about 1.4 million new cases of cancer would be diagnosed, and 500,000 patients would die of it in the United States alone.[1-3] Although major advances have been achieved in our understanding of cancer biology and pathogenesis as well as in the development of new-targeted therapies, the progress in developing improved early diagnosis and screening tests has been inadequate. As a result, most cancers are diagnosed in advanced stages delaying timely treatment and leading to poor outcomes.

There is intense research seeking specific molecular changes that are able to identify patients with early cancer or precursor lesions. Biological samples like blood, serum, stool, pancreatic juice, urine etc., as well as both DNA and RNA have been analysed for tumour-specific changes. Additionally, genomic DNA alterations, circulating viral DNA or RNA, various mutations such as K-RAS, p16 and/or APC mutations either in serum, blood or circulating cancer cells in blood samples have been evaluated to allow the diagnosis of cancer patients. Recently, epigenetic changes such as gene methylation in biological samples from cancer patients have also been analysed. In addition, serum levels of certain proteins involved in tumour biology, like cathepsin B, E-cadherin, hepatocyte growth factor, interleukins, and other cytokines and hormones, have been measured in the serum of cancer patients. However, none of these analysis methods have yet shown adequate sensitivities and specificities in order to facilitate the detection of cancer in its early stages.[4]

Recently, a novel class of global gene regulators called microRNAs (miRNAs), small non-coding single-stranded RNAs of about 22 nucleotides, were identified in both plants and animals.[5] MiRNAs regulate gene expression in two ways; first, miRNAs that bind to protein-coding mRNA sequences that are exactly complementary to the miRNA induce the

RNA-mediated interference (RNAi) pathway, leading to cleavage of mRNA by ribonucleases in the RNA-induced silencing complex (RISC).[6–8] In the second and more common mechanism, miRNAs exert their effect by binding to imperfect complementary sites within the 3′ untranslated regions (3′ UTRs) of their target protein-coding mRNAs, leading to repression of expression of these genes at the level of translation.[9–15] Consistent with translational control, miRNAs can reduce protein levels of their target genes with minimal impact on the genes' mRNAs. Accumulating and encouraging evidence demonstrates the importance of miRNAs in cancer. MicroRNAs that are overexpressed in cancer may function as oncogenes, and miRNAs with tumour suppressor activity in normal tissue may be downregulated in cancer.[16]

9.1.1 *MicroRNA profiling in identifying early malignant changes*

The vast amount of microRNA expression data from a large number of independent studies in various cancers have shown that cancer cells have different miRNA profiles compared with normal cells thus underscoring the tremendous diagnostic potential of miRNAs in cancer. A pioneer study by Lu *et al.* has very successfully demonstrated this. The study reports that 129 of 217 microRNAs were expressed to a lower extent in all tumour samples compared to the normal tissue. They also showed that expression profiles of only a few hundred microRNAs could correctly classify the tumours originating from different tissues. Their microRNA classifier clustered the tumours from endothelial lineage like colon, liver, pancreas and stomach together, and tumours of hematopoietic origin were clustered as a separate group. On the contrary, profiling of more than 15,000 mRNA genes failed to group the tumours accurately.[17]

One example of the power of miRNAs in differentiating normal from tumour tissue is evidence that microRNA expression patterns can successfully separate pancreatic cancer from benign pancreatic tissue; 21 miRNAs are differentially overexpressed and 4 are underexpressed relative to chronic pancreatitis and adjacent benign pancreas tissue, indicating the diagnostic role of miRNAs in pancreatic cancer. Several studies found that miR-196a levels were significantly increased in pancreatic cancer tissue compared with normal tissue as well as normal pancreatic lines and acute pancreatitis specimens. MiR-217 exhibited opposite expression patterns, suggesting its potential diagnostic power.[18,19] Another study by Bloomston *et al.* has demonstrated that aberrant expression of 22 miRNAs could accurately distinguish benign and pancreatic ductal adenocarcinomas with 93% accuracy.[20]

A study by Volinia *et al.* was the first to show that 26 miRNAs were overexpressed and 17 downregulated in six kinds of solid cancers including stomach.[21] Subsequent studies by different groups demonstrated several microRNAs are associated with gastric cancer. MiR-21, miR-20b, miR-20a, miR-17-5p, miR-106a, miR-18a, miR-106b, miR-18b, miR-421, miR-340*, miR-19a and miR-658 are all overexpressed in gastric cancer compared to adjacent non-tumourous tissue. Findings by the Chan group demonstrated that miR-21 is overexpressed in 92% of gastric cancer, and those by Xiao *et al.* showed 1.625-fold increased expression of miR-106a in 55 gastric carcinoma samples compared to 17 normal samples, and this was also confirmed in gastric cancer cell lines.[22–24] These results are suggestive of potential diagnostic roles for miR-21 and miR-106a in gastric cancer.

Another elaborate genome-wide miRNA expression profiling study by Calin's group reports the robust downregulation of miR-15a and miR-16 in 75% of chronic lymphocyte leukaemia (CLL) cases. MiR-15a and miR-16 are located on chromosome 13q14, a region frequently deleted in CLL.[25,26] As opposed to this the miR-17-92 polycistron, encoded within the 13q31, a region that is commonly amplified in B-cell lymphoma, was upregulated in 65% of the B-cell lymphoma patients.[27]

9.1.2 *Differentiating tumour subtypes*

Over 50% of the lung cancer cases are diagnosed as adenocarcinomas (AC) and squamous cell carcinomas (SCC). However, both are categorised as non-small cell cancer and although treated similarly the prognosis is different. One study reported that in four out of five AD and two out of eight SCC, miR-34b was found downregulated by more than 90%.[28] Several independent studies also reported a markedly higher expression of miR-34b, miR-34c and miR-449 in normal lines and normal lung, compared to lung cancer lines and tissue as well as any other normal tissue.[17,29,30] Liang *et al.*

studied the predicted target genes of miR-34b/34c/449 using computational algorithms and found that expression of 17 genes could more accurately classify AD and SCC in independent data sets than the histological diagnosis. Furthermore, the expression of the same gene signature of the bronchial epithelial cells from cigarette smokers in combination with cytopathology of the bronchial cells successfully discriminated between patients with or without lung cancer, thus demonstrating the diagnostic power of microRNAs and their targets in lung cancer.[31] A study by Yanaihara *et al.* found 43 miRNAs were differentially expressed between normal and different subtypes of lung cancer tissues, moreover this miRNA signature correctly classified 91% of 104 lung cancers, even in Stage I cancer.[32] The *let-7* miRNA is also frequently significantly downregulated in 80% of lung cancers versus corresponding normal lung tissues. All these results indicate that miRNA expression profiles are diagnostic markers of subtypes of lung cancer.[33]

Data from several studies has depicted the strong association of deregulated miRNA expression with breast cancer. In 2005, Iorio *et al.* carried out a genome-wide miRNA expression profiling of a large set of breast cancer and normal tissue. Twenty-nine miRNAs were found to be differentially expressed in breast cancer versus normal tissues, miR-21 and miR-155 were upregulated, whereas miR-10b, miR-125b and miR-145 were downregulated, suggesting that these miRNAs may potentially act as diagnostic markers. Subsequent studies showed that expression patterns of miR-21 and miR-145 could discriminate between cancer and normal tissues.[34,35]

Another group has demonstrated the potential of miR-145 as a novel biomarker for breast cancer diagnosis.[36] A bead-based flow cytometric method established by Blenkiron *et al.* showed that miRNA profiling of 93 primary human breast tumours accurately classified those as Luminal A, Luminal B, Basal-Like, HER2+ and normal-like. Particularly they found that miR-155 could discriminate ER- and ER+ tumours.[37] In addition, a study led by Scott *et al.* profiling microRNAs in a cohort of 20 different breast tumours, followed by supervised analysis, identified a unique subset of miRNAs that distinguished HER2+ from HER2− and ER+ from ER- breast cancers.[38] Findings from all these studies suggest that miRNAs possibly influence the pathogenesis of different cancer subtypes and hence function as effective diagnostic markers differentiating them.

9.1.3 MicroRNAs in detecting early cancers

The development of biomarkers that help detect cancer at an early stage is important since early detection has a direct impact on prognosis and clinical outcome as evidenced by a higher (49%) five-year survival in lung cancer patients diagnosed at an early stage compared to those diagnosed later (15%). Currently, employed tests for tumour biomarkers are cumbersome, time consuming, labour intensive and offer a relatively limited number of targets. Considering the simplicity and minimal invasive nature, the development of blood, serum or plasma biomarkers is of considerable value.

The small size, relative stability and resistance to RNase degradation make the miRs more superior molecular markers than mRNAs.[39] The recent advances in qRT-PCR methods have improved the sensitivity of miRNA detection to a few nanograms of total RNA thus making it possible to quantify miRNA by qRT-PCR on fine-needle aspiration biopsy samples.[40] Furthermore, a couple of recent studies have demonstrated that miRNAs are extremely stable in human plasma and serum where they are protected from RNAses. Their stability and predictive properties make them ideal candidates to be tested in patient serum and plasma samples. This provides a means of direct measurement of the miRNAs from patients' blood surpassing the need for invasive procedures. A report by Chen *et al.* showed that miRNA expression profiles from serum of healthy individuals was significantly different from that of patients with lung cancer, colorectal cancer and diabetes.[41]

Another very recent study by the Gao Research Group found profound overexpression of 5 miRNAs both in plasma and tissue samples of colorectal cancer (CRC) patients compared to that in healthy controls. Of these miR-17-3p and miR-92 were significantly elevated in CRC patients ($p < 0.0005$). Furthermore, these researchers demonstrated that miR-92 correctly discriminated CRC from gastric cancer, IBD and normal subjects in an independent set of plasma samples.[42] Colorectal cancer (CRC) is the third most common cancer worldwide. The current colonoscopic screening is both invasive as well as costly while the fecal occult blood test is limited by low sensitivity and meticulous dietary restriction. Therefore the finding that miR-92 can serve as a potential non-invasive molecular marker for CRC screening is very encouraging.

Several other studies have shown the diagnostic potential of miRs that can be measured in serum. Lawrie *et al.* demonstrated that miR-21 levels are high in the serum from patients with diffuse B-cell lymphoma and this was associated with relapse-free survival.[43] Also, studies in mice implanted with human prostate cancer cells demonstrated high circulating levels of tumour-derived microRNAs. High serum miR-141 levels were detected in metastatic prostate cancer patients and were also found to identify prostate cancer patients with high accuracy.[44] In a study conducted by Chen *et al.*, unique serum microRNA profiles were identified in patients with lung cancer, colorectal cancer and diabetes. Furthermore, 63 miRNAs not present in healthy individuals were observed in the serum from lung cancer patients. Specifically, miR-25 and miR-223 with high expression in the sera from lung cancer patients were found to be biomarkers for non-small cell lung cancer.[41]

An interesting study by Taylor *et al.* profiled miRNAs in circulating tumour-derived exosomes in ovarian cancer. Exosomes are small (50–100 nm) membrane vesicles of endocytic origin, in the peripheral circulation of women with ovarian cancer. This group found that levels of the eight specific microRNAs were similar between cellular and exosomal microRNAs. These exosomal microRNAs displayed similar profiles in ovarian cancer patients but were significantly distinct from those in benign disease, and exosomal microRNAs were not found in normal controls.[45] These results suggest that microRNA profiling of circulating tumour exosomes can serve as a potential diagnostic tool in ovarian cancer.

9.1.4 *MicroRNAs in diagnosing unknown tumour origin*

A large number of tumour profiling studies in different cancers have shown that each cancer and normal tissue has a unique microRNA signature that distinguishes normal from neoplastic tissue, premalignant lesions from malignant ones, and primary tumours from one organ system to another. Moreover, miRNAs show differential expression across different tumour types.[46] Additionally, the repression or overexpression of certain miRNAs has been correlated with aggressive or metastatic phenotypes. Metastatic cancers of unknown primary origin represent a unique class with a frustrating diagnosis and huge treatment challenge for the patient as well as the oncologist. Over the past number of years, cancer chemotherapy has become more targeted according to tumour type and primary site of origin. Cancers of unknown primary origin are defined as histologically-confirmed metastatic tumours for which no known primary site has been identified. Autopsy studies have revealed the lungs, pancreas, liver, kidney, gut, bone etc. to be the possible primary sites.[47] These cancers of unknown primary represent approximately 3–6% of aggressive malignancies, ranking among the 10 most common malignancies with poor prognosis and ill-defined therapeutic strategies due to a lack of evidence-based options. In addition, previous attempts to better classify them and determine their tissue of origin by conventional mRNA-based expression profiling have failed. As a result all cancer types in this category tend to be treated equivalently without consideration for the cell type of the origin, leading to reduced antineoplastic efficacy and poor outcome.

Using microRNA-based classification, Rosenfeld has recently attempted to define tumour identity in tumours of unknown origin. MiRNA microarrays run on 22 different tumour tissues and metastases were used to construct a classifier based on 48 miRNAs. This classifier, when used on a blinded test set of 83 samples, predicted tissue type with approximately 90% accuracy. The classification system developed in these studies could therefore be utilised to identify the tissue of origin in cancers of unknown primaries. Moreover, despite the smaller number of miRNAs (a few hundred) than the protein-coding genes (several thousands), the unique yet differential miRNA expression patterns correlated more accurately with cancer type, stage and clinico-pathological variables than gene profiling.[48]

Similarly, a study by Lu *et al.* profiled miRNA expression in 17 poorly-differentiated tumours with non-diagnostic histological appearance and demonstrated that their smaller miR classifier made a superior diagnosis of the samples compared to that by a large mRNA classifier. This miRNA classifier was also found to be more informative with better predictive power for cancer of unknown primary (CUP) diagnosis than the traditional profiling of several thousands of mRNA genes.[17] This could be a significant step towards the use of miRNA-based classification and identification of cancers of unknown origin — a major clinical problem — and can pave the way to the use of more personalised and targeted therapeutic strategies.

9.1.5 *MicroRNAs in diagnosing cancer predisposition*

The notion that single nucleotide polymorphisms (SNPs) in protein-coding genes can affect the functions of proteins and in turn influence the individual susceptibility to cancers has been well documented. However, the role of miRNA-associated SNPs in disease is just emerging. Because small variations in the quantity of miRNAs may have an effect on thousands of target mRNAs and result in diverse functional consequences, the most common genetic variation, SNPs, in miRNA sequences may also be functional and therefore may represent ideal candidate biomarkers for cancer diagnosis, prognosis and outcome. SNPs that disrupt miRNA-coding sequences have been associated with cancer risk. Inherited mutations or rare SNPs in the primary transcripts of *hsa-mir-15a* and *hsa-mir-16-1* have been linked to familial chronic lymphocytic leukaemia and familial breast cancer.[26] Several miRNA-associated SNPs have been shown to increase breast cancer susceptibility, for example, a SNP, rs11614913, located in the 3p mature miRNA region, has been identified in the miRNA *has-mir-196a2*. The variant genotypes CC/CT were associated with significantly increased breast cancer risks in a case-control study of 1,009 breast cancer cases and 1,093 cancer-free controls in a population of Chinese women. Similarly, the subjects carrying variant homozygous genotypes hsa-mir-499 rs3746444:A > G displayed significantly increased risks of breast cancer (OR, 1.75; 95% CI, 1.07–2.85 for rs3746444 GG, respectively) compared with their wild-type homozygotes.[49] Another study conducted by Shen *et al.* has identified a G to C polymorphism (rs2910164) within the sequence of *miR-146a* precursor and demonstrated that variant C allele led to increased levels of mature miR-146 in breast and ovarian cancer patients, and predisposed them to an earlier age of onset of familial breast and ovarian cancer.[50] All of these findings suggest, for the first time, that common SNPs in miRNAs may contribute to breast cancer susceptibility and may serve as novel biomarkers for breast cancer diagnosis.

On the other hand, in a cancer association study of 479 hepatocellular carcinoma (HCC) and 504 control subjects, Xu *et al.* demonstrated that male individuals with GG genotype in rs2910164 were 2-fold more susceptible to HCC (odds ratio = 2.016, 95% confidence interval = 1.056 – 3.848, $P = 0.034$) compared with those with CC genotype.[51] Furthermore, in an association study of case control for parathyroid carcinoma (PTC), Jazdzewski *et al.* found that individuals' heterozygous (GC genotype) for the SNP had an increased risk of acquiring PTC (OR = 1.62, 95% CI = 1.3 – 2.0, *P* = 0.000007).[52]

A recent study genotyping 40 SNPs from 11 miRNA processing genes and 15 miRNA genes in 279 Caucasian patients with renal cell carcinoma and 278 matched controls, reported that 2 SNPs in the GEMIN4 gene were significantly associated with altered renal cell carcinoma risks. This indicates a possible putative role for the genetic polymorphisms of the miRNA-machinery genes in the diagnosis of renal cell carcinoma.[53]

The first evidence that miRNAs may affect oesophageal cancer risk in general, and that the specific genetic variants in miRNA-related genes may affect oesophageal cancer risk individually and jointly, comes from the studies conducted by Wu *et al.* in a case-control study of 346 Caucasian oesophageal cancers. They found that seven SNPs were significantly associated with oesophageal cancer risk, the most prominent being the homozygous wild-type genotype of the SNP rs6505162 located in the pre-mir423 region.[54]

Two other SNPs in the miRNA processing pathway genes XPO5 and RAN were also identified and associated with an increased oesophageal cancer risk. The XPO5 SNP has also been reported to be associated with an increased risk of renal cell carcinoma.[53,54]

There is also evidence that miRNA binding site SNPs can influence cancer risk. Two recent papers report SNPs in miRNA target sites in human cancer genes[55] and show that allele frequencies vary between normal and cancerous tissue.[56] Furthermore, a SNP identified in miRNA binding site in the *kit* oncogene was associated with increased gene expression in papillary thyroid carcinoma.[57] Another SNP has been identified in a *let-7* binding site in the KRAS oncogene, which disrupts *let-7* regulation of KRAS and is also associated with altered cellular miRNA levels. This SNP (LCS6SNP) has been shown to be a biomarker of an increased risk to developing non-small cell lung cancer (NSCLC) in two independent case-control studies.[58]

Taken together, the findings from all these studies demonstrate the valuable utility of microRNA-associated SNP evaluation in cancer diagnosis.

9.2 MicroRNAs in Cancer Therapeutics

9.2.1 *Ectopic expression of miRNAs as cancer therapy*

9.2.1.1 *MiRNAs as tumour suppressors*

There are a number of microRNAs that are downregulated in cancers. For example, miR-125b-1, the homologue of *C. elegans* lin-4, is deleted in a subset of patients with breast, lung, ovarian and cervical cancers.[59] In addition to serving as a possible biomarker, some of these downregulated miRNAs are implicated as tumour suppressors. There is a significant reduction in miR-143 and miR-145 in colorectal tumours as well as in breast, prostate, cervical and lymphoid cancer cell lines. Wang *et al.* demonstrated that cervical cancer cell lines and cervical cancer tissue consistently had loss of miR-143, miR-145 expression whereas normal cervical cells or cervical tissue did not.[60] These two miRNAs are putative tumour suppressors; functionally, the addition of both of these miRNAs to cell lines inhibited proliferation.[59]

Mechanistically, their functional roles are explained by their ability to downregulate genes that promote cancer pathogenesis, including oncogenes, anti-apoptotic genes and stem cell maintenance genes. The *let-7* family of miRNAs represents a classic example. The *let-7* family of miRNAs encodes 12 human homologues. They are present at a fragile site associated with lung, breast, urothelial and cervical cancers.[33] Underscoring their role as tumour suppressors, they were shown to be downregulated in patients with non-small cell lung cancer; furthermore, the downregulation of *let-7* correlated with a poor prognosis.[61]

Regarding the mechanism, this family of miRNAs were the first group of oncomirs that was shown to regulate the expression of an oncogene, namely the *RAS* genes.[62] The *RAS* proteins are important membrane-associated GTPases implicated in about 15–30% of human tumours. Normally associated with cellular growth and differentiation, activating mutations of *RAS* are thought to support several hallmarks of cancer, including independence from growth factors, evasion of apoptosis, angiogenesis and metastasis/invasion. The 3′ UTR of *RAS* genes contains multiple complementarity sites for the *let-7* family. In addition, *RAS* and *let-7* are reciprocally expressed when comparing tissue from patients with lung tumours to those with adjacent normal tissue. Functionally, ectopic *let-7* expression downregulated *RAS* expression, and inhibitors of *let-7* reversed this inhibition.[33] Ultimately, *let-7* expression was shown to inhibit cellular proliferation *in vitro* tissue-culture experiments with a human lung adenoma cell line.[26]

MiR-15a and miR-16 are two clustered miRNA genes that are found in 13q14. Deletions of 13q14 occur in 65% of chronic lymphocytic leukaemia (CLL) cases, as well as in 50% of mantle cell lymphomas, 16–40% of multiple myelomas and 60% of prostate cancers.[26] Cimmino *et al.* showed that miR-15a and miR-16-1 may increase cell death by negatively regulating Bcl-2, a known antiapoptotic gene overexpressed in many types of human cancers.[63] MiR-15a and miR-16 expression inversely correlated with Bcl-2 protein levels in samples from CLL patients. The 3′ UTR of Bcl-2 contained a conserved target site for miR-15a and miR-16. Furthermore, overexpression of these microRNAs led to decreased bcl2 protein expression and ultimately apoptosis.[48]

Irrespective of cell type, more than half of the miRNAs are downregulated in tumours compared with normal tissues. This may reflect a role of miRNAs in terminal differentiation. Underscoring this possibility, differentiation of primary human haematopoietic stem cells along the erythroid lineage induced many miRNAs; similarly, addition of a differentiation factor, all-trans retinoic acid also induced expression of many miRNAs in a myeloid leukaemia cell line.[64] MiR124 and miR137 appear to be two such miRNAs. They have been found to be downregulated in glioblastoma multiforme and anaplastic astrocytoma. Interestingly, they are also relatively downregulated in neuronal stem cells. Conditions conducive to differentiation, i.e. growth factor withdrawal, induce miR-124 and miR-137 expression. Furthermore, ectopic expression of these miRNAs leads to upregulation of differentiation markers, i.e. Tuj1, and downregulation of markers of stem cells, i.e. GFAP. Underscoring their functional role in cancer, ectopic expression of these miRNAs in CD133+ tumour stem cells reduces cell proliferation.[64]

9.2.1.2 *Restoring miRNA expression in vivo*

Because of the strong association between the miRNA *let-7* and lung cancer, it appears to have potential as a therapeutic agent. To explore this possibility, Esquela-Kercher *et al.* tested whether ectopic expression can reduce tumour loads *in vitro* and *in vivo*.[27] They showed that transfection of lung cancer cell lines with let-7b inhibited proliferation; these included cell lines with and without activating mutations in KRAS. To determine their potential role *in vivo*, the group transfected lung cancer cell lines in a xenograft mouse experiment. In this model, lung cancer cell lines are injected subcutaneously into the flanks of mice; pre-implantation transfection with let-7b significantly reduced tumour volume compared to transfection with a control miRNA. To further probe its potential efficacy, the group also used let-7b in an orthotopic model of murine lung cancer. In this model, KRAS G12D contains a stop codon floxed within it. Addition of a cre-recombinase via adenovirus activates the gene and leads to tumour formation. Intranasal treatment of these mice with adenovirus containing both the Cre gene as well as let-7 miRNA led to a 90% reduction in lung hyperplasia compared to treatment with adenovirus containing the Cre gene alone.[65] In this example, using a LacZ reporter, they were able to demonstrate stable expression of the adenovirus containing the miRNA in the bronchial tree for up to two weeks. It remains unclear whether such *let-7* will treat human lung cancer. The pharmacokinetics remain unclear; the extent of penetration with the adenovirus and the stability of *let-7* expression remain to be explored in humans.

9.2.2 *MiRNA antagonism as anti-cancer therapy*

9.2.2.1 *MicroRNAs as oncogenes*

MicroRNAs that promote proliferative or antiapoptotic activity would likely promote oncogenesis, be overexpressed in cancer cell lines and represent potential targets for therapy. The targets for the oncogenic microRNAs, when identified, have been putative tumour suppressors. Antisense oligonucleotides have been shown to consistently be able to inhibit miRNA *in vitro*, upregulate expression of miRNA gene targets, and ultimately reverse miRNA dependent phenotypes. Detailed below are several examples of potential miRNA targets for antagonistic therapy.

The most comprehensively studied example consists of the miR-17-92 cluster, which includes seven miRNAs: miR-17-5p, miR-17-3p, miR-18a, miR-19a, miR-20a, miR-19b1 and miR-92-1. The 13q31 locus, which harbours these genes, is preferentially amplified in cancers such as diffuse large B-cell lymphoma, follicular lymphoma, mantle cell lymphoma and primary cutaneous B-cell lymphoma. This cluster is upregulated in 65% of B-cell lymphoma samples tested, in B-cell CLL, as well as in lung cancers and soft tissue tumours.[66] In a mouse model, Eμ-*Myc* transgenic mice develop B-cell lymphomas relatively late in life and with incomplete penetrance. Coexpression of a truncated form of the miR-17-92 cluster leads to a dramatically accelerated onset of disease.[66] In contrast to Eμ-*Myc* lymphomas, which have many apoptotic cells, miRNA-expressing lymphomas have a high mitotic index without extensive apoptosis. One potential mechanism is attributed to modulation of a pro-proliferative/pro-apoptotic transcription factor E2F1.[67]

Another miRNA with apparent oncogenic activity is miR-21. It is strongly upregulated in breast cancer samples;[68] upregulated 5–100-fold in glioblastoma cell lines and tissue;[69] and upregulated over 9-fold in hepatocellular carcinoma cell lines.[70] Underscoring its functional significance, antagonism of miR-21 with antisense oligonucleotides leads to increased caspase activation and subsequent apoptosis in glioma cell lines.[71] In hepatocellular carcinoma (HCC) cell lines, antagonism leads to decreased proliferation. Furthermore, there is decreased migration across a transwell with multiple effects along the invasion phenotype with downregulation of metalloproteinases and reciprocal activation of focal adhesion kinases. Putative gene targets of miR-21 include phosphatase and tensin homologue (PTEN) tumour suppressor; in HCC lines, overexpression of PTEN partially reverses miR-21's oncogenic effects.[72]

MiR-155 is another likely oncogene. It has been found to be upregulated in a number of cancers. MiR-155 is found at a common integration for the avian leukosis virus and found to be overexpressed in B-cell lymphomas, specifically in cells that overexpress the *Myc* oncogene. MiR-155 is upregulated 100-fold in pediatric Burkitt lymphoma.[73] There is similar upregulation of miR-155 in the Hodgkin lymphoma, in primary mediastinal and some diffuse large B-cell lymphomas.[74] MiR-155 has also been shown to be upregulated in breast carcinomas as well as cervical cancers.[74] Underscoring its oncogenic potential, upregulation of miR-155 in haematopoietic stem cells has led to myeloid neoplasms in a mouse model.[75]

9.2.2.2 *Antagomirs*

Krutzfeldt *et al.* demonstrated that microRNAs can be silenced *in vivo*. They designed chemically-engineered oligonucleotides called 'antagomirs'. These moieties conjugate oligonucleotides with a sequence to silence endogenous miRNAs with a hydroxyproline-linked cholesterol solid support and 2'-OMed phsopharmidites. They showed that these antagomirs can target miR-122 in the liver whereas chemically-modified but unconjugated single-stranded RNAs could not. Further, they showed that antagomirs can target endogenous miRNAs everywhere outside the brain, including the liver, kidney, lung, skin and muscle.[69] Presumably, this *in vivo* efficacy is due to stability.

In 2008, antagomirs were used *in vitro* and *in vivo* in a model of therapy-resistant neuroblastoma. Fontana *et al.* demonstrated that neuroblastomas had upregulation of a gene MYCN. This particular gene, like *c-Myc*, upregulates a cluster of miRNAs in the 17-5p-92 cluster. In *in vitro* and *in vivo* assays, the 17-5p miRNA is sufficient to mediate the effects of MYCN in particular increases in proliferation, growth on soft-agar, and in a xenograft mouse model. Of particular note, antagomirs to 17-5p miRNA were able to reverse the MYCN phenotype *in vitro* and *in vivo*. In the xenograft mouse model, injection of antagomir 17-5p specifically prevented increase in tumour growth, increased markers of apoptosis and increased expression of two putative 17-5p targets: the tumour suppressors p21 and BIM (Bcl-2-interacting-mediator of cell death).[70]

9.2.3 *MiRNAs as adjuvants to conventional therapy*

9.2.3.1 *MiRNAs as modulators of radiation therapy*

In a search for miRNAs that modulate radiation response, Weidhaas *et al.* showed that levels of 81 miRNAs significantly changed after irradiation. The patterns of miRNA expression changes in response to radiation were comparable in the lung cancer cell line and the untransformed cell line. Interestingly, all members of the *let-7* family were significantly different before and after radiation treatment. All of the *let-7* miRNAs were significantly downregulated except for let-7g. Underscoring the functional significance of these miRNAs, overexpression of let-7b — a miRNA downregulated by radiation — appeared to increase radiosensitivity. Downregulating let-7b with anti-let-7b anti-miRNAs decreased cell death in response to radiation. In a parallel fashion, overexpression of let-7g, a miRNA normally upregulated by radiation, increased radioprotection. Addition of anti-let-7g led to increased radiation dose-dependent cell death. The similarity in the response of cancer cells and untransformed cells suggested that there is a highly-conserved global miRNA response to irradiation in lung cells. As the *let-7* data illustrate, these may reflect components of a protective cellular response to cytotoxic DNA damage, and manipulation of this response may enhance radiation-killing.[71]

9.2.3.2 *MiRNAs as modulators of chemotherapy response*

Investigators have examined drug-resistant cancer cell lines, searching for genes that mediate drug resistance. In examination of breast cancer cell lines, they have found several potential miRNA targets. Miller *et al.* found a number of miRNAs that are differentially expressed between a parental tamoxifen-sensitive breast cancer cell line and its tamoxifen resistant derivative.[72] Fifteen miRNAs were differentially regulated. Two of these miRNAs (miR-221, miR-222) were also found to be upregulated in Her2/neu+ breast cancer tumour samples, which are known to be tamoxifen resistant. Overexpression of the miRNAs is sufficient to confer resistance to tamoxifen-induced apoptosis. A possible mechanism that has been postulated is the regulation of the p27(Kip1) protein which is downregulated by miR-221/miR-222. A similar study with a breast cancer cell line with a doxorubicin-resistant derivative, suggested the possible involvement of miR-451, which can modulate expression of the multidrug resistance MDR1 protein.

In cholangiocarcinomas, miRNA targets were identified that may sensitise these otherwise highly chemoresistant tumours to chemotherapy. In human cholangiocyte cell lines, Meng *et al.* demonstrated that a subset of miRNA specifically upregulated in cholangiocarcinoma cell lines may alter the response to chemotherapy. MiR-21 and miR-200b were among 21 that were selectively upregulated in cancer cell lines but not in untransformed controls. In a xenograft *in vivo* model, addition of gemcitabine increased expression of these miRNAs. Anti-miR-21 and anti-miR-200b both increased gemcitabine-induced cell death in these cell lines, in part by increasing chemotherapy-dependent apoptosis. Likewise, addition of these miRNAs in the nonmalignant cholangiocyte cell line increased cell viability in response to gemcitabine.[76] A possible mechanism proposed was the modulation of tumour suppressors, PTPN12 and PTEN.

9.2.4 *MicroRNAs in metastasis: potential for future miRNA therapy*

Metastasis is thought to be responsible for 90% of cancer-related deaths. The mechanisms by which cancers metastasize remain unclear. Furthermore, it is unclear if this represents a process in carcinogenesis that can be targeted and prevented. Nevertheless, recent papers have demonstrated that subsets of miRNAs may suppress metastasis in breast cancer and others may enhance metastasis. Tavazoie *et al.* showed that there are endogenous miRNAs that appear to inhibit metastasis.[74] There were six microRNAs that were downregulated in human breast cancer cell derivatives that were highly metastatic to bone or lung. Six miRNAs — miR-335, miR-126, miR-206, miR-122a, miR-199a* and miR-489 — were most downregulated in metastatic cell lines compared to parental, unselected populations. When human tumour cells were isolated from metastatic lesions that formed in the lungs and bones of the mice, they were found to have significantly reduced levels of miR-335, miR-206 and miR-126. Restoring the expression of miR-335 and miR-206 in LM2 cell lines altered cell morphology and reduced transwell migration *in vitro* and reduced lung colonising activity *in vivo*. Conversely, targeting miR-335 with an anti-miRNA antagomir enhanced the lung-colonising ability of MDA-MB-231 cells compared with control cells. Further underscoring its role in metastasis, miR-335 appears to reduce the expression of genes that have already been shown to be highly expressed in metastatic breast cancer cells. MiR-335, miR-206 and miR-126 appear to have specific effects on metastasis as they had no effect on proliferation or apoptosis. Additional work showed that the miR-200 family of microRNAs suppresses epithelial-mesenchymal transition (a process implicated in metastasis) via modulation of the transcription factors ZEB1 and ZIP1. Dysregulation of these miRNAs has been corroborated in a subset of invasive breast cancer cell lines, namely those with a mesenchymal phenotype.

Conversely, Ma *et al.* showed that miR-10b can augment metastasis. They found that miR-10b is specifically upregulated in metastatic breast cancer cell lines. Transfection of miR-10b antagonists in metastatic breast cancer cell lines reduced *in vitro* transwell migration. Reciprocally, overexpression of miR-10b enhanced transwell migration of non-metastatic cell lines. Ectopic expression of miR-10b in an *in vivo* xenograft model of metastasis increased locally invasive features, including invasion into stroma, vessels and muscle. In addition, such expression led to distant metastases in the lung. Huang, *et al.* performed a screen looking for miRNAs that cause migration of breast cancer cell lines across a transwell. They found that miR-373 and miR-520c can also increase metastases in a xenograft *in vivo* model, and miR-373 is upregulated in metastatic breast cancer cell lines.[75]

Conclusions

The current diagnostic methods in cancer lack sufficient sensitivity and specificity to facilitate the detection of cancer in its early stages. Someday, profiling blood miRNAs as a diagnostic test for cancer would be a huge advance in the field, and a non-invasive as well as an easy alternative. Diagnosis of metastatic cancers with an unknown primary site of origin is very frustrating both for the patient and for the physician, and also poses huge treatment challenges. The power of miRNAs in accurately identifying the tissue of origin in such cases is a significant advance. The study of miRNA SNPs and miRNA-binding site SNPs as biomarkers of cancer risk is another way in which miRNAs may open up new avenues, allowing cancer prevention. By identifying those at greatest risk, the application of improved diagnostic and screening tests becomes more cost-effective.

The evidence implicating miRNA deregulation in the initiation and progression of cancer has opened up new opportunities for exploiting the miRNA system for therapeutic manipulations and development of novel therapies. As a result, oncogenic miRNAs can be targeted for downregulation using anti-sense miRNA oligonucleotides and tumour suppressive miRNAs may be directly upregulated for an anti-cancer effect. With the rapid technological advances and our understanding of microRNA biology in the near future, miRNAs may be administered in cancer therapeutics either as single agents or in combination therapies. Studies have already shown that miRNAs can play a direct role in drug effectiveness. Delivery systems such as locked nucleic acid increase the stability of miRNAs for *in vivo* application and may be promising as cancer therapy.

Taken together, the studies reviewed here clearly demonstrate that several aspects of miRNAs, including their intricate nature of interaction with multiple targets and multiple pathways, make them extremely useful potential agents for clinical diagnostics as well as novel therapeutics in personalised care for individual patients in the future.

Acknowledgments

Trupti Paranjape was supported by CTSA Grant Number UL1 RR024139 from the National Center for Research Resources (NCRR), a component of the National Institutes of Health (NIH) and NIH roadmap for Medical Research. Its contents are solely the responsibility of the authors and do not necessarily represent the official view of NCRR or NIH. JBW was supported by a NIH K08 grant (CA124484) and a grant from the NIH (CA131301-01A1).

References

1. American Cancer Society. (2006). Cancer statistics 2006. [Online]. Available at: http://www.cancer.org
2. Dalton W.S. and Friend S.H. (2006). Cancer biomarkers — An invitation to the table. *Science* 312, 1165–1168.
3. Mathers C.D. and Loncar D. (2006). Projections of global mortality and burden of disease from 2002 to 2030. *PLoS Med* 3, e442.
4. Ebert M.P., Korc M., Malfertheiner P. *et al.* (2006). Advances, challenges, and limitations in serum-proteome-based cancer diagnosis. *J Proteome Res* 5, 19–25.
5. Bartel D.P. (2004). MicroRNAs: Genomics, biogenesis, mechanism, and function. *Cell* 116, 281–297.
6. Hamilton A.J. and Baulcombe D.C. (1999). A species of small antisense RNA in posttranscriptional gene silencing in plants. *Science* 286, 950–952.
7. Reinhart B.J., Weinstein E.G., Rhoades M.W. *et al.* (2002). MicroRNAs in plants. *Genes Dev* 16, 1616–1626.
8. Yekta S., Shih I.H. and Bartel D.P. (2004). MicroRNA-directed cleavage of HOXB8 mRNA. *Science* 304, 594–596.

9. Abrahante J.E., Daul A.L., Li M. *et al.* (2003). The *Caenorhabditis elegans* hunchback-like gene lin-57/hbl-1 controls developmental time and is regulated by microRNAs. *Dev Cell* **4**, 625–637.

10. Lee R.C., Feinbaum R.L. and Ambros V. (1993). The *C. elegans* heterochronic gene lin-4 encodes small RNAs with antisense complementarity to lin-14. *Cell* **75**, 843–854.

11. Lin S.Y., Johnson S.M., Abraham M. *et al.* (2003). The *C. elegans* hunchback homolog, hbl-1, controls temporal patterning and is a probable microRNA target. *Dev Cell* **4**, 639–650.

12. Moss E.G., Lee R.C. and Ambros V. (1997). The cold shock domain protein LIN-28 controls developmental timing in *C. elegans* and is regulated by the lin-4 RNA. *Cell* **88**, 637–646.

13. Olsen P.H. and Ambros V. (1999). The lin-4 regulatory RNA controls developmental timing in *Caenorhabditis elegans* by blocking LIN-14 protein synthesis after the initiation of translation. *Dev Biol* **216**, 671–680.

14. Reinhart B.J., Slack F.J., Basson M. *et al.* (2000). The 21-nucleotide let-7 RNA regulates developmental timing in *Caenorhabditis elegans*. *Nature* **403**, 901–906.

15. Slack F.J., Basson M., Liu Z. *et al.* (2000). The lin-41 RBCC gene acts in the *C. elegans* heterochronic pathway between the let-7 regulatory RNA and the LIN-29 transcription factor. *Mol Cell* **5**, 659–669.

16. Zhang B., Pan X., Cobb G.P. *et al.* (2007). MicroRNAs as oncogenes and tumor suppressors. *Dev Biol* **302**, 1–12.

17. Lu J., Getz G., Miska E.A. *et al.* (2005). MicroRNA expression profiles classify human cancers. *Nature* **435**, 834–838.

18. Lee E.J., Gusev Y., Jiang J. *et al.* (2007). Expression profiling identifies microRNA signature in pancreatic cancer. *Int J Cancer* **120**, 1046–1054.

19. Szafranska A.E., Davison T.S., John J. *et al.* (2007). MicroRNA expression alterations are linked to tumorigenesis and non-neoplastic processes in pancreatic ductal adenocarcinoma. *Oncogene* **26**, 4442–4452.

20. Bloomston M., Frankel W.L., Petrocca F. *et al.* (2007). MicroRNA expression patterns to differentiate pancreatic adenocarcinoma from normal pancreas and chronic pancreatitis. *JAMA* **297**, 1901–1908.

21. Volinia S., Calin G.A., Liu C.G. *et al.* (2006). A microRNA expression signature of human solid tumors defines cancer gene targets. *Proc Natl Acad Sci USA* **103**, 2257–2261.

22. Guo J., Miao Y., Xiao B. *et al.* (2008). Differential expression of microRNA species in human gastric cancer versus non-tumorous tissues. *J Gastroenterol Hepatol.* **24**, 652–657.

23. Schetter A.J., Leung S.Y., Sohn J.J. *et al.* (2008). MicroRNA expression profiles associated with prognosis and therapeutic outcome in colon adenocarcinoma. *JAMA* **299**, 425–436.
24. Xiao B., Guo J., Miao Y. *et al.* (2009). Detection of miR-106a in gastric carcinoma and its clinical significance. *Clin Chim Acta* **400**, 97–102.
25. Calin G.A., Dumitru C.D., Shimizu M. *et al.* (2002). Frequent deletions and down-regulation of micro- RNA genes miR15 and miR16 at 13q14 in chronic lymphocytic leukemia. *Proc Natl Acad Sci USA* **99**, 15524–15529.
26. Calin G.A., Ferracin M., Cimmino A. *et al.* (2005). A microRNA signature associated with prognosis and progression in chronic lymphocytic leukemia. *N Engl J Med* **353**, 1793–1801.
27. He L., Thomson J.M., Hemann M.T. *et al.* (2005). A microRNA polycistron as a potential human oncogene. *Nature* **435**, 828–833.
28. Liang Y., Ridzon D., Wong L. *et al.* (2007). Characterization of microRNA expression profiles in normal human tissues. *BMC Genomics* **8**, 166.
29. Bommer G.T., Gerin I., Feng Y. *et al.* (2007). p53-mediated activation of miRNA34 candidate tumor-suppressor genes. *Curr Biol* **17**, 1298–1307.
30. Dutta K.K., Zhong Y., Liu Y.T. *et al.* (2007). Association of microRNA-34a overexpression with proliferation is cell type-dependent. *Cancer Sci* **98**, 1845–1852.
31. Liang Y. (2008). An expression meta-analysis of predicted microRNA targets identifies a diagnostic signature for lung cancer. *BMC Genomics* **1**, 61.
32. Yanaihara N., Caplen N., Bowman E. *et al.* (2006). Unique microRNA molecular profiles in lung cancer diagnosis and prognosis. *Cancer Cell* **9**, 189–198.
33. Takamizawa J., Konishi H., Yanagisawa K. *et al.* (2004). Reduced expression of the let-7 microRNAs in human lung cancers in association with shortened postoperative survival. *Cancer Res* **64**, 3753–3756.
34. Iorio M.V., Casalini P., Tagliabue E. *et al.* (2008). MicroRNA profiling as a tool to understand prognosis, therapy response and resistance in breast cancer. *Eur J Cancer* **44**, 2753–2759.
35. Iorio M.V., Ferracin M., Liu C.G. *et al.* (2005). MicroRNA gene expression deregulation in human breast cancer. *Cancer Res* **65**, 7065–7070.
36. Sempere L.F., Christensen M., Silahtaroglu A. *et al.* (2007). Altered microRNA expression confined to specific epithelial cell subpopulations in breast cancer. *Cancer Res* **67**, 11612–11620.
37. Blenkiron C., Goldstein L.D., Thorne N.P. *et al.* (2007). MicroRNA expression profiling of human breast cancer identifies new markers of tumor subtype. *Genome Biol* **8**, R214.

38. Mattie M.D., Benz C.C., Bowers J. *et al.* (2006). Optimized high-throughput microRNA expression profiling provides novel biomarker assessment of clinical prostate and breast cancer biopsies. *Mol Cancer* **5**, 24.

39. Waldman S.A. and Terzic A. (2007). Translating microRNA discovery into clinical biomarkers in cancer. *JAMA* **297**, 1923–1925.

40. Pallante P., Visone R., Ferracin M. *et al.* (2006). MicroRNA deregulation in human thyroid papillary carcinomas. *Endocr Relat Cancer* **13**, 497–508.

41. Chen X., Ba Y., Ma L. *et al.* (2008). Characterization of microRNAs in serum: A novel class of biomarkers for diagnosis of cancer and other diseases. *Cell Res* **18**, 997–1006.

42. Ng E.K., Chong W.W., Jin H. *et al.* (2009). Differential expression of microRNAs in plasma of colorectal cancer patients: A potential marker for colorectal cancer screening. *Gut* **58**, 1375–1381.

43. Lawrie C.H., Gal S., Dunlop H.M. *et al.* (2008). Detection of elevated levels of tumour-associated microRNAs in serum of patients with diffuse large B-cell lymphoma. *Br J Haematol* **141**, 672–675.

44. Mitchell P.S., Parkin R.K., Kroh E.M. *et al.* (2008). Circulating microRNAs as stable blood-based markers for cancer detection. *Proc Natl Acad Sci USA* **105**, 10513–10518.

45. Taylor D.D. and Gercel-Taylor C. (2008). MicroRNA signatures of tumor-derived exosomes as diagnostic biomarkers of ovarian cancer. *Gynecol Oncol* **110**, 13–21.

46. Nelson K.M. and Weiss G.J. (2008). MicroRNAs and cancer: Past, present, and potential future. *Mol Cancer Ther* **7**, 3655–3660.

47. Pentheroudakis G., Golfinopoulos V. and Pavlidis N. (2007). Switching benchmarks in cancer of unknown primary: From autopsy to microarray. *Eur J Cancer* **43**, 2026–2036.

48. Rosenfeld N., Aharonov R., Meiri E. *et al.* (2008). MicroRNAs accurately identify cancer tissue origin. *Nat Biotechnol* **26**, 462–469.

49. Hu Z., Liang J., Wang Z. *et al.* (2009). Common genetic variants in pre-microRNAs were associated with increased risk of breast cancer in Chinese women. *Hum Mutat* **30**, 79–84.

50. Shen J., Ambrosone C.B., DiCioccio R.A. *et al.* (2008). A functional polymorphism in the miR-146a gene and age of familial breast/ovarian cancer diagnosis. *Carcinogenesis* **29**, 1963–1966.

51. Xu T., Zhu Y., Wei Q.K. *et al.* (2008). A functional polymorphism in the miR-146a gene is associated with the risk for hepatocellular carcinoma. *Carcinogenesis* **29**, 2126–2131.

52. Jazdzewski K., Murray E.L., Franssila K. *et al.* (2008). Common SNP in pre-miR-146a decreases mature miR expression and predisposes to papillary thyroid carcinoma. *Proc Natl Acad Sci USA* **105**, 7269–7274.

53. Horikawa Y., Wood C.G., Yang H. *et al.* (2008). Single nucleotide polymorphisms of microRNA machinery genes modify the risk of renal cell carcinoma. *Clin Cancer Res* **14**, 7956–7962.

54. Ye Y., Wang K.K., Gu J. *et al.* (2008). Genetic variations in microRNA-related genes are novel susceptibility loci for esophageal cancer risk. *Cancer Prev Res* **1**, 460–469.

55. Yu Z., Li Z., Jolicoeur N. *et al.* (2007). Aberrant allele frequencies of the SNPs located in microRNA target sites are potentially associated with human cancers. *Nucl Acids Res* **35**, 4535–4541.

56. Landi D., Gemignani F., Barale R. *et al.* (2008). A catalog of polymorphisms falling in microRNA-binding regions of cancer genes. *DNA Cell Biol* **27**, 35–43.

57. He H., Jazdzewski K., Li W. *et al.* (2005). The role of microRNA genes in papillary thyroid carcinoma. *Proc Natl Acad Sci USA* **102**, 19075–19080.

58. Chin L.J., Ratner E., Leng S. *et al.* (2008). A SNP in a let-7 microRNA complementary site in the KRAS 3' untranslated region increases non-small cell lung cancer risk. *Cancer Res* **68**, 8535–8540.

59. Calin G.A., Sevignani C., Dumitru C.D. *et al.* (2004). Human microRNA genes are frequently located at fragile sites and genomic regions involved in cancers. *Proc Natl Acad Sci USA* **101**, 2999–3004.

60. Wang X., Tang S., Le S. *et al.* (2008). Aberrant expression of oncogenic and tumor-suppressive microRNAs in cervical cancer is required for cancer cell growth. *PLoS ONE* **3**, e2557.

61. Esquela-Kerscher A., Trang P., Wiggins J. *et al.* (2008). The let-7 microRNA reduces tumor growth in mouse models of lung cancer. *Cell Cycle* **7**, 759–764.

62. Johnson S., Grosshans H., Shingara J. *et al.* (2005). RAS is regulated by the let-7 microRNA family. *Cell* **120**, 635–647.

63. Hwang H. and Mendell J. (2006). MicroRNAs in cell proliferation, cell death, and tumorigenesis. *Br J Cancer* **94**, 776–780.

64. Esquela-Kerscher A. and Slack F. (2006). Oncomirs — MicroRNAs with a role in cancer. *Nat Rev Cancer* **6**, 259–269.

65. O'Donnell K., Wentzel E., Zeller K. *et al.* (2005). c-Myc-regulated microRNAs modulate E2F1 expression. *Nature* **435**, 839–843.

66. Chan J.A., Krichevsky A.M. and Kosik K.S. (2005). MicroRNA-21 is an anti-apoptotic factor in human glioblastoma cells. *Cancer Res* **65**, 6029–6033.

67. Metzler M., Wilda M., Busch K. *et al.* (2004). High expression of precursor microRNA-155/BIC RNA in children with Burkitt lymphoma. *Gene Chromosome Canc* **39**, 167–169.
68. O'Connell R., Rao D., Chaudhuri A. *et al.* (2008). Sustained expression of microRNA-155 in hematopoietic stem cells causes a myeloproliferative disorder. *J Exp Med* **205**, 585–594.
69. Krützfeldt J., Rajewsky N., Braich R. *et al.* (2005). Silencing of microRNAs *in vivo* with 'antagomirs'. *Nature* **438**, 685–689.
70. Fontana L., Fiori M., Albini S. *et al.* (2008). Antagomir-17-5p abolishes the growth of therapy-resistant neuroblastoma through p21 and BIM. *PLoS ONE* **3**, e2236.
71. Weidhaas J., Babar I., Nallur S. *et al.* (2007). MicroRNAs as potential agents to alter resistance to cytotoxic anticancer therapy. *Cancer Res* **67**, 11111–11116.
72. Miller T., Ghoshal K., Ramaswamy B. *et al.* (2008). MicroRNA-221/222 confers tamoxifen resistance in breast cancer by targeting p27(Kip1). *J Biol Chem* **283**, 29897–29903.
73. Kovalchuk O., Filkowski J., Meservy J. *et al.* (2008). Involvement of microRNA-451 in resistance of the MCF-7 breast cancer cells to chemotherapeutic drug doxorubicin. *Mol Cancer Ther* **7**, 2152–2159.
74. Tavazoie S., Alarcón C., Oskarsson T. *et al.* (2008). Endogenous human microRNAs that suppress breast cancer metastasis. *Nature* **451**, 147–152.
75. Ma L., Teruya-Feldstein J. and Weinberg R. (2007). Tumour invasion and metastasis initiated by microRNA-10b in breast cancer. *Nature* **449**, 682–688.
76. Meng F., Henson R., Lang M. *et al.* (2006). Involvement of human micro-RNA in growth and response to chemotherapy in human cholangiocarcinoma cell lines. *Gastroenterology* **130**, 2113–2129.

10

MicroRNA Target Prediction

Isidore Rigoutsos*,‡ and Aristotelis Tsirigos†

Over the last few years, researchers have witnessed an explosion in the number of identifiable classes of short RNAs that appear to be functionally distinct. Among them, microRNAs represent the best-studied group to date. Despite the enormous research effort that has been devoted to the study of microRNAs, many questions remain open including what is arguably the most important of all: given a microRNA, what are its targets? In what follows, we discuss and summarise the key aspects of the various schemes that have been devised so far for answering this question.

10.1 Introduction

MicroRNAs are a class of short RNAs, approximately 22 nucleotides in length, which can affect the expression of their targets in a manner that depends on the underlying sequence.[1] The first member of the class, *lin-4*, was discovered in 1993[2,3] in the context of studies involving developmental timing in *C. elegans*. Following this first discovery, nearly seven years went by before the second member of the class, *let-7*, was reported[4], again in *C. elegans*. Despite this slow start, soon thereafter the microRNA class grew in size by leaps and bounds as new members were discovered in plants,[5] other animals[6] and eventually in viruses.[7,8]

---

*Department of Pathology, Anatomy and Cell Biology, Jefferson Medical College, Thomas Jefferson University, Philadelphia, Pennsylvania 19107, USA. †IBM TJ Watson Research Center, Yorktown Heights, NY 10598, USA. ‡E-mail: isidore.rigoutsos@jefferson.edu.

The discovery of microRNAs very quickly raised the questions of the identity and cardinality of their targets. For the human genome, early estimates placed the average number of targets for a given microRNA to a few tens, with approximately 25–30% of the known human genes thought to be under microRNA control.[9] As the number of discovered microRNAs increased so did the need for an automated method for discovering their targets.

Abstractly, the task at hand can be summarised as follows: given the sequence of a microRNA *m* and a genome of interest, identify the genes and the location(s) within the genes' transcripts that are targeted by *m*. The original discoveries had demonstrated that *lin-4* and *let-7* acted on their targets by forming heteroduplexes with the 3′ untranslated region (3′ UTR) of the corresponding gene. These early observations drove nearly all research efforts for many years by implicitly delineating the region of the mRNA where microRNA targets were to be sought. The topic of how the adherence to 3′ UTRs as recipients of microRNA targeting activity shaped both experimental and computational approaches for many years, was recently discussed elsewhere[10] and will not be revisited here. The rest of this discussion will focus on the various methodologies that were proposed over the course of the last several years for tackling the microRNA target prediction problem. Given the limited amount of available space, we will only describe and discuss the most representative of those schemes.

10.2 Definitions and Background

The special sequence relationship between a microRNA's 5′ end and that of its targets has been pointed out repeatedly over time and was understood more clearly after the recent reporting of the crystal structure of a RNA:DNA heteroduplex.[11] Indeed, at the time of the initial discovery of the *lin-4:lin-14* interaction, it was noted that *lin-14*'s 3′ UTR contained seven sequence segments that matched perfectly — or nearly so — the reverse complement of *lin-4*'s 5′ region;[3] this region of the microRNA was initially referred to as the 'core element'.

The reported *lin-4:lin-14* interactions made clear that the heteroduplexes fell into two groups: those whose target sites contained the exact reverse complement of the core element and those that contained a near-instance of it.[3] The latter group comprised target sites with either G:U wobbles in the core element region of the interaction, or bulges (i.e. unpaired bases). Additional work examined these 'non-canonical' sites in the context of both *in vitro* and *in vivo* experiments, presumably in an effort to determine their relevance, and found them to be essential for generating a temporal gradient during *C. elegans* development.[12] Analogous results were reported in the context of *let-7*'s control of the timing of developmental events.[4,13]

Effectively, the arguments supporting the presence of wobbles and bulges in the core element region were as many as those supporting their absence. However, in those early days no rules were apparent with regard to the permitted number and location of these deviations from normal base pairing. There is universal agreement among researchers that the 5' region of the microRNA is indeed very important for the interaction with its targets. However, this importance has been interpreted and used in different ways giving rise to seeming contradictions among the various contributions to the literature.

More than a decade after its original discovery, the 'core element' began being referred to as the 'seed',[14] a label that has been used pervasively ever since. In terms of actual location, accumulating evidence helped delineate the seed between positions 2 and 7 inclusive counting from the 5' end of the microRNA; this was referred to as the 6-mer seed. In some instances, the seed could be augmented by incorporation of up to two flanking nucleotides: addition of either the nucleotide at position 8 or the nucleotide at position 1 (typically an A) would form 7-mer seeds, whereas simultaneous addition of both would form 8-mer seeds.[15]

By the time the concept of the seed became well defined, a good number of heteroduplexes had been reported in the literature. Careful studies revealed that the heteroduplexes fell into three distinct classes that could be defined based on the spatial arrangement of their interactions.[16] The first class comprised 'canonical' heteroduplexes that were defined by strong basepairing in both the seed region and the 3' end of the microRNA. The heteroduplexes of the second class were characterised by basepairing that spanned only the seed region of the microRNA; experimental evidence suggested that seed-only target sites were more effective in suppressing their target if the latter contained more than one instance of the seed's

Figure 10.1. The three types of heteroduplexes. Left: canonical. Middle: 5′-dominant. Right: 3′-compensatory. The seed is delineated by the dotted rectangle. Adapted from Ref. 16.

reverse complement spaced close to one another, or if they contained 7-mer or 8-mer seeds. Finally, the third class was characterised by weaker basepairing in the seed region that was supported by pronounced basepairing in the 3′ region of the microRNA, hence the name '3′ compensatory'. These three classes are summarised in Figure 10.1.

10.3 Methods

In this section, we present and discuss in some detail the more representative methods that have been devised for discovering microRNA targets. We have not always adhered to the chronological ordering in which the various methods appeared in the literature. Instead, we have grouped them primarily based on the key ideas that underlie them. This allows for a thematic separation while at the same time permits the comparison of methods that are based on similar principles.

10.3.1 'Seed'-driven methods

TargetScan[9] represents one of the earliest attempts to employ the seed concept in determining microRNA targets. Once presented with the microRNA of interest, TargetScan seeks instances of the seed's *exact* reverse complement (i.e. no bulges or wobbles are permitted in the seed region) in the 3′ UTRs of putative target genes. It is important to note that only orthologous 3′ UTR regions are considered in the search for target sites and that the latter are expected to be conserved across organisms. Candidate sites are extended through incorporation of additional basepairs immediately beyond and contiguous to the seed: for locations beyond the seed, wobbles are allowed but extension ceases at the first mismatch. The block of contiguous basepairs formed up to this point is used as an anchor in conjunction with the RNAFold structure prediction programme[17] to form a possibly extended heteroduplex that involves basepairs along the length of the microRNA (see Figure 10.1). These heteroduplexes are then scored using a scheme that favours multiple sites within the 3′ UTR for the same microRNA and ranked in order of decreasing score. Heteroduplexes that receive scores above a predetermined threshold are reported to the user. The original version of TargetScan was evaluated using a HeLa cell reporter assay that validated 11 of 15 tested predictions. This result in conjunction with shuffled sequence controls allowed the authors to estimate the rate of false positives of the algorithm to approximately 30% of the predictions. Many of the ideas that underlie TargetScan are also present in MovingTargets, which appeared soon thereafter.[18]

MiRanda[19] is similar in spirit to TargetScan but has several noteworthy differences. Most important among them is the fact that, in acknowledgement of the literature examples involving *lin-4*'s interactions with its targets, miRanda permits G:U wobbles in the seed region as well as a limited number of mismatches (but forces them to be beyond position 4 of the microRNA). In terms of implementation, it uses a modified version of a classical dynamic programming solution to the problem of string editing,[20] which itself is a variant of an even earlier-published scheme.[21] Use of dynamic programming allows miRanda to naturally determine putative heteroduplexes that can extend to the full length of the microRNA. The authors capture many of the various constraints that they employ (e.g. the concept of a seed region, G:U wobbles, bulges, etc.) in the scoring matrix that they use in their dynamic programming approach. MiRanda was originally developed for finding microRNA targets in *D. melanogaster* but subsequently modified to find targets in human 3′ UTRs.[22] Both instances of miRanda constrain the search for microRNA targets to only orthologous 3′ UTRs, just like TargetScan does. We will revisit the conservation constraint and its uses below. We also note that the concept of using a scoring matrix in the search for putative microRNA target sites was revisited in the Stacking Binding Matrix (SBM) approach;[23] this method relies on validated microRNA targets to build a multiple sequence alignment and subsequently a scoring matrix, which is bound to introduce biases and does not generalise to cases for which no representative examples are available.

With respect to enforcing the strict presence of the seed within a predicted target site, the method presented by Stark *et al.*[24] is positioned between TargetScan and miRanda: it does permit the presence of G:U wobbles in the seed region but not of bulges. For the microRNA of interest the method requires the building of two hidden Markov models using the HMMer package.[25] The first model contains only copies of the exact reverse complement of positions 1–8 inclusive of the microRNA. The second model extends the first by adding instances of the reverse complement with C and A replaced by T and G respectively in order to permit G:U wobbles. Both models are then used to search 3′ UTRs for putative target sites of the microRNA at hand. The discovered 3′ UTR hits are extended in their 5′ direction and the resulting sequence is paired with the microRNA and evaluated using Mfold, another RNA structure prediction programme.[26,27] The free energy of the formed heteroduplex is compared to that of a background distribution and a Z-score determined: the heteroduplexes whose Z-scores exceed threshold are kept and reported. The method reduces the number of false positives by an estimated 5-fold by only considering 3′ UTR regions that are conserved across *D. melanogaster* and *D. pseudoobscura*. Additional filtering is performed by the optional requirement that the targets also be conserved in *A. gambiae*.

Another algorithm that is very similar to the three that we have already described was reported in.[28] The algorithm is based on the observation, which the authors made from the experimentally validated heteroduplexes that were available at the time, that there is a preponderance of G:C bonds in short segments of the microRNA that basepair with the target. The authors used the term 'nuclei' to refer to those regions. The method employs the sequence composition of the nucleus to identify candidate target sites but weighs differently each basepair in the nucleus region, scoring G:C bonds higher than A:T bonds and allowing no G:U wobbles. Even though, as the authors mention, their analyses show all such nuclei were found very close to the 5′ end of the microRNA, in the general case they do not have to be. The algorithm proceeds in two phases. Initially, and for a given microRNA, candidate target sites are identified by seeking putative such nuclei in the 3′ UTRs of the genome of interest. Each instance is scored using different weights for the basepairs in the nucleus and the total score compared to a background distribution of scores that is generated offline. Those of the candidate nuclei that exceed a predefined threshold are extended in both their 5′ and 3′ directions to generate a corresponding 3′ UTR segment approximately 40 nts in length that is then paired with the microRNA and evaluated using Mfold.[26,27] Only those targets whose free energy is less than half of the free energy that would have been obtained by pairing the full-length microRNA with its exact reverse complement are kept and reported. An interesting novel element of this algorithm is that although it also relies on the cross-genomic presence of predicted microRNA targets in the 3′ UTRs of orthologous genes, it does not require that these targets be *syntenically* conserved in these genes.

10.3.2 *Non-seed driven methods*

RNAhybrid represents the chronologically first reported method that does not require that the seed be present in the sequence of a putative target.[29] The method uses a dynamic programming scheme to compute minimum free energies for all possible relative arrangements between a microRNA and a putative target with the required energy contributions provided by earlier work.[26] It deviates from the seed-driven algorithms in that the requirement of a seed-region basepairing is now optional. In fact, the algorithm allows the user to arbitrarily define a region of perfect, Watson-Crick basepairing anywhere along the length of the microRNA, which is in some respect reminiscent of the nucleus-driven approach described in earlier work.[28] The resulting putative heteroduplexes are then evaluated with the help of an extreme value distribution of length-normalised free energies and reported if they exceed a predefined threshold.

A method that is rather distinct from the ones that we have discussed so far is rna22.[30] Probably the most important characteristic of rna22 is that it does not make use of validated heteroduplexes for its training phase. As a matter of fact, it does not use any of them, relying instead on the available repository of mature microRNAs.[31] The key idea underlying the rna22 approach is the following: if there exist salient sequence features that are shared by the currently known mature microRNAs, whether the microRNAs belong to the same or to different organisms, these features should be easily identifiable using a pattern discovery technique;[32] more-over, the reverse complement of these patterns ought to be able to capture salient sequence features that are shared by microRNA target sites. This last observation should in principle allow for the identification of microRNA target sites as regions where such patterns cluster making these sites distinguishable from the rest of the sequence. The authors used a previously published combinatorial, exhaustive pattern discovery algorithm[33] to discover several hundred thousand sequence motifs shared by various subsets of the mature microRNAs in Release 4.0 of miRBase.[31] As is shown by Miranda et al.,[30] this key idea holds for nearly all validated target sites that were available at the time. After putative microRNA target sites are identified in the above manner, the method makes use of RNAfold[17] to determine the matching microRNA as the one that forms an energetically-favourable heteroduplex with the extracted target sites. A similar in spirit method but rather limited in the kinds of patterns that it explored and the manner in which it explored them was TargetBoost.[34]

Since it does not make use of validated heteroduplexes, rna22 does not exhibit any 3′ UTR bias. This allowed the authors to explore other parts of the mRNA sequence and to argue computationally that 5′ UTRs and CDS regions were equally feasible targets of microRNAs.[30] In the case of CDS regions, several papers recently showed this to be the case.[35–39]

Because rna22 first identifies putative microRNA target sites in a candidate mRNA sequence, and then determines the identity of the targeting microRNA, it does not make use of the seed in the manner that the other methods do. This in turn allows the algorithm to report functioning heteroduplexes containing one or more bulges and/or G:U wobbles in the seed region; several such examples are shown in several publications.[30,39] Finally, this method neither requires nor uses any cross-genome conservation constraints for the microRNA targets it reports; this permitted the discovery and validation of several targets that are not conserved across genomes.[37,39,40]

Three more observations that were made by Miranda et al.[30] are of relevance to this review. The first pertains to the number of genes potentially under microRNA control: using rna22's predictions and pretty stringent thresholds, the authors reported their estimate that more than 90% of the human genes are controlled by microRNAs. The second observation relates to it, is also derived computationally, and states that a given microRNA may have as many as a few thousand targets. Both of these estimates appear to be corroborated by more recent findings.[41,42] The third observation relates to a modification of the target prediction algorithm that Miranda *et al.* also presented[30] and allows the discovery of microRNA precursors and of their mature microRNAs: using this modification of rna22, the authors predicted that the eventual numbers of microRNAs to be found in the genomes of model organisms like human, mouse, fruitfly and the worm, are substantially higher than all previous estimates — see Table 10.2 of Miranda *et al.*[30] The number of microRNAs that are harboured in an organism's genome continues to remain a hotly-contested topic.

10.3.3 *Conservation requirement and sequence-context*

As described already, and as we will see again below in the discussion of other methods for predicting microRNA targets, the requirement that the sequence of a predicted microRNA target site be conserved across organisms has been a very pervasive one. The underlying assumption here is that similarity among sequences that are separated by large evolutionary distances is likely to be indicative of positive selection and thus be of functional relevance. The original observation was made more than 25 years ago[43] and has since fueled very wide-ranging research activities along similar lines.[44,45] Researchers in the microRNA field made analogous findings simultaneously with the early discoveries.[3,12] However, it was not until much later that such findings were put in use in the context of microRNA target selection.

The practical reason for using the cross-genome conservation constraint is the resulting improvement of the signal-to-noise ratio and the increased confidence in the putative biological significance of the reported predictions. By enforcing the constraint across different species or different genera the signal improves and so does the ability to localise the sequence of interest. However, this improved ability comes at the expense of a concomitant decrease in sensitivity.

In the context of cross-genome conservation, we also wish to mention the work described by Xie *et al.*[46] Even though, strictly speaking, the authors do not describe a microRNA target detection method, they provide a very good example of the rationale behind the use of this constraint. In particular, the authors aligned the human, mouse, rat and dog genomes focusing on promoter regions and 3′ UTRs. They identified conserved regions that they conjectured correspond to regulatory motifs. The motif collection that they derived from the promoter regions contained an abundance of known transcription factor binding sites as well as novel binding motifs. On the other hand, analysis of the 3′ UTRs yielded several tens of conserved 8-mer motifs: the reverse complement of many of these motifs were matched to microRNAs that existed in the publicly available databases at the time; the rest of the motifs were linked to genomic sequences that folded into microRNA-precursor-like secondary structures and experiments with a small, representative sample showed that as many as 50% could be novel microRNAs. The manner in which known and novel microRNAs were linked to known 3′ UTRs shares key elements with many of the existing methods for microRNA target detection (but was carried out in the reverse manner). In a more recent effort, the same basic approach was repeated on 12 species of the Drosophilidae family with analogous results.[47]

Recently, the findings of the ENCODE project suggested that sequence conservation is neither a necessary nor a sufficient condition for functional significance.[48] Similar findings were also reported by other groups not connected with the ENCODE project.[49–52] The immediate implication of these observations for microRNA targeting is that enforcing the cross-genome conservation constraint is bound to miss *bona fide* targets and thus underestimates their number. A number of publications that appeared very recently provide yet another a strong argument in support of this point.[10,37,39,40]

Even though helpful in improving the signal-to-noise ratio in computationally predicted targets, the cross-genome conservation constraint is not always effective in weeding out poor predictions. One recurrent observation has been that the degree of target repression depends on context. This leads to the incorporation of a number of additional filters, derived from the study of validated heteroduplexes, which attempt to capture this dependence by examining the sequence surrounding the predicted target site. Probably the most important among these filters is the examination of the sequence in the immediate vicinity of a predicted target for additional such sites either for the same or for a different microRNA.[15,53] MicroRNA target sites that are spaced reasonably close to one another appear to be more effective than sites further apart.

Another observation was that heteroduplexes which include basepairing spanning the third quarter of a microRNA's sequence lead to more effective downregulation of their target — see Figure 10.1 and discussion by Brennecke *et al.*[16] A filter that has been used by several groups is the measure of the A–T richness of the sequence immediately surrounding the predicted target site. Sites surrounded by A–T rich segments are more preferred to those that are not, as A–T richness correlates with lack of structure increasing the chance that the site is exposed and available for binding by the microRNA. Over time, this filter has been used by practitioners implicitly[54–56] as well as explicitly.[15] Finally, the position of a microRNA target within the 3' UTR was argued to be a useful criterion: sites near the two ends were reported to be more effective than those elsewhere along the length of the 3' UTR;[15] however, exceptions do exist.[57] Even though these sequence-based filtering schemes proved effective in identifying biologically significant microRNA target sites, it should be stressed that they ought to be viewed as general guidelines. As a matter of fact, the continued reporting in the literature of validated heteroduplexes suggests that there are many exceptions to these empirical rules.

In addition to sequence filters, several other types were also proposed and incorporated into the various tools. We examine them separately in the next sub-section.

10.3.4 *Incorporating non-sequence information*

DIANA-microT[58] goes beyond the validated heteroduplexes that were available at the time. First, the authors used a dynamic programming approach to identify putative targets for a microRNA by sliding a 38-nucleotide window along the length of available 3' UTR sequences in hops of 3 nucleotides; G:U wobbles and bulges in the seed region were permitted. A group of identified heteroduplexes involving putative target sites that were conserved in human and mouse was selected and subjected to experimental testing using reporter assays. Additional heteroduplexes corresponding to mutated variants of the predicted targets were also tested with reporter assays. The results of these experiments permitted the authors to determine a set of rules that were then embedded in the target prediction tool to assist in filtering out non-promising targets.

The PicTar algorithm[59] takes a slightly different approach in an effort to improve the signal-to-noise ratio in the generated target predictions. In particular, it uses information from microRNA profiling studies to identify co-expressed microRNAs which it then uses to determine the likelihood that a given sequence is bound by a *combination* of microRNAs. The core module that searches for an individual microRNA's putative target sites is the one described in Rajewsky and Socci,[28] extended to permit bulges in the seed region but not G:U wobbles. Predicted sites are filtered further using free energy considerations as well as through the application of cross-species conservation constraints, the latter being considered by the authors a crucial component in ensuring low rates of false-positives.

Similarly to PicTar, GenMiR++ allows for multiple microRNAs per target gene and multiple binding sites per microRNA.[60] In addition to considering microRNA expression information, GenMiR++ takes into account mRNA expression that it models as a function of microRNA expression and attempts to find the optimal assignment of microRNAs to target genes so as to maximise the likelihood of the available expression data. GenMiR++ is a supervised method: indeed, it uses a set of known targets genes to train its model. However, considering the potentially large diversity in microRNA targets, the available validated examples are few by comparison; thus, their use in training is expected to introduce biases. Another very important assumption that underlies the GenMiR++ approach is that microRNA activity manifests itself at the mRNA level as a result of mRNA degradation and thus can be accurately captured by expression data. This has been a contested topic, but there is increasing evidence suggesting that animal microRNAs may act primarily through translational inhibition.[35–37,39,56,61] It should be pointed out that, strictly speaking, GenMiR++ is not a microRNA target prediction tool: instead it is an extension that permits the scoring of targets that have been predicted by other tools and the identification of those predictions that are likely to be functionally meaningful.

Even though the focus of the work presented by Zhao *et al.*[56] was on miR-1 and not on microRNA target prediction, we have chosen to include it in this review because the authors introduced a novel and interesting filter for sub-selecting among computationally predicted targets. Using the sequence of miR-1, they searched for putative targets using the reverse complement of miR-1's seed as a proxy and enforcing a cross-genome conservation requirement (across human, mouse, rat and chicken). This search is like TargetScan with the exception that a G:U wobble is allowed at position 8 of the microRNA. The novel filter that was introduced was the evaluation of the free energy of the two 70-nucleotide stretches on the 5′ and 3′ sides of the predicted target: targets that are surrounded by *destabilising* neighbourhoods are favoured as they imply an increased lack of structure, structural accessibility and, by extension, enhanced target prediction specificity. The methodology was demonstrated in the context of cardiogenesis by showing that miR-1 targets Hand2, a transcription factor that promotes ventricular cardiomyocyte expansion. The same basic concept was also discussed by Robins *et al.*,[55] but in that work the authors confined validation to the use of reporter assays only, unlike the *in vivo* validation experiments presented by Zhao *et al.*[56]

More recently, the same idea was revisited in PITA.[54] Strictly speaking, PITA is a seed-driven approach to predicting microRNA targets. During its first stage, PITA identifies putative targets for a given microRNA by seeking instances of the reverse complement of the seed, effectively mirroring TargetScan. Similarly to the work described by Zhao *et al.*,[56] PITA permits at most one G:U wobble in the seed region, but the wobble can be located anywhere in the seed region. PITA deviates from TargetScan and the work by Zhao *et al.*[56] by dismissing the cross-genome conservation requirement. Once a candidate microRNA target has been identified, it is flanked by 70 nucleotides on each side, the resulting neighbourhood extracted and its free energy computed without and with the microRNA bound to it. Those heteroduplexes for which the computed energy difference satisfies a predetermined threshold, the latter having been derived with the help of previously validated heteroduplexes, are assumed to represent predicted targets that are sterically accessible by the microRNA, and, thus are kept and reported.

More recently, Hammell *et al.*[62] described mirWIP, an approach that uses immunoprecipitation-derived mRNA targets of microRNAs to deduce principles characterising functional heteroduplexes. These principles are combined with a target prediction method in order to help rank the putative targets. In that respect, it is reminiscent of the work by Kiriakidou *et al.*[58] The authors focused on the following dimensions: (a) structural accessibility of the putative target sequences; (b) total free energy of miRNA-target hybridization; and (c) topology of base-pairing to the 5′ seed region of the miRNA. The authors use a modified version of RNAhybrid to generate putative microRNA target sites that are then scored using the derived rules.[29] The rules that the authors deduce from the immunoprecipitation-derived data can, of course, be incorporated into any of the other available target prediction tools and provide a ranking for generated results.

We conclude this section by mentioning two schemes based on machine learning: miTarget[63] and NBmiRTar.[64] We include them in this section because they both rely on features that go beyond the available sequence information. The first uses a support vector machine approach whereas the second relies on a naïve Bayes classifier. The two schemes share many similarities but NBmiRTar represents a substantial extension over MiTarget in that it uses a much richer and rather diverse set of features. Both schemes require positive and negative examples for training purposes. In order to cope with the relative small number of negative examples that are in the public domain, the groups had to devise novel approaches for generating artificial examples of negative data points, and this may be introducing biases in the performance of the tools. The different ways in which the tools were evaluated in comparison to earlier methods makes a direct comparison between them difficult. Finally, it should be noted that NBmiRTar is more general than miTarget in that it can be used as an add-on filter to any of the other target prediction tools.

10.3.5 *Non-computational methods for identifying microRNA targets*

The advent of experimental methods exhibiting increased accuracy and efficiency has generated new opportunities for designing procedures that can either directly identify microRNA targets, or that can be used as a first filter whose output is subsequently exploited by a computational scheme. These methods are still in their infancy and, naturally, they are limited by the speed at which the experimental procedures can be carried out.

Vatolin *et al.*[65] present an experimental method to directly identify miRNA–mRNA complexes. During a first stage, they use mature microRNAs as primers to generate reverse-transcribed molecules. These are purified and used as templates with gene-specific primers during a second round of reverse transcription. Sequencing of the generated cDNA and a database search can reveal the composition of the mRNA that was originally targeted and allows the elucidation for part of the sequence of the targeting microRNA. The recovery of only short microRNA stretches limits the usefulness of the method as there is no statistical test to verify whether these stretches could have arisen by chance.

In the work of Easow *et al.*,[66] miRNP immunopurification is proposed as a scheme that permits the direct identification of microRNA targets. The proposed method was validated in only a limited manner using a single microRNA, miR-1 and samples from Drosophila. The presentation includes several very interesting observations not the least of which is the finding that only approximately 30% of the immunopurified targets of the studied microRNAs contained the reverse complement of the seed in their 3' UTRs. There are at least two, not mutually exclusive, explanations: the microRNA target sites may be located outside the 3' UTR of the identified targeted gene or the principles governing the interaction of a microRNA and its target go beyond the presence of the seed's reverse complement in a mRNA. However, neither these nor other possibilities were explored at length by the authors. An essentially identical method is described by Karginov *et al.*[67] In the latter, the authors focused on one microRNA, miR-124a, in an effort to facilitate comparison with the microarray-derived results that were described earlier by Lim *et al.*[68]

Recently, two different groups reported experimental results from using the SILAC method[69,70] to determine the impact that the induction (respectively knockdown) of a microRNA can have on protein synthesis.[41,42] The questions that the two groups set out to answer were the same and included: determining whether the seed is necessary or sufficient in determining a gene target; resolving the importance of the 3' UTR in the context of microRNA targeting; answering the question of how many simultaneous direct targets a microRNA can have; determining whether the primary mode of action of a microRNA is through degradation or inhibition, etc. The findings by the two groups to a certain degree recapitulated the findings of earlier reports[14] but notable differences were also observed. First, it became clear that among the genes that were directly repressed by the studied microRNAs, only a relatively small fraction contained the microRNA's seed in their 3′ UTRs, thus still leaving open the question of what rules govern the interaction between microRNAs and mRNAs. Secondly, both groups established that each microRNA can simultaneously target directly several hundreds of genes, in concordance with earlier theoretical predictions.[30] Thirdly, with regards to whether microRNA activity results in mRNA degradation or translational inhibition, one group found degradation to represent the most significant component of the observed repression[41] whereas the other found equally compelling evidence in support of both modes.[42] Interestingly, the results also show that there were genes whose 3′ UTRs contained the reverse complement of the seed of the microRNA at hand and which were not affected, in agreement with earlier reports of a specific microRNA:mRNA interaction from *C. elegans*.[57] Finally, both groups found evidence for microRNA targeting outside 3′ UTRs but considered such targets as less effective by comparison to their 3′ UTR counterparts, in direct contrast with the findings reported elsewhere.[35–37,39,40]

Notwithstanding the results in Baek *et al.* and Selbach *et al.*,[41,42] a word of caution is warranted. The reporting of these two groups is derived from the analysis of only a very small number of microRNAs and is based on a method that is relatively insensitive by comparison to other approaches. Moreover, in order to ensure a reasonable degree of confidence, both groups employed thresholds that were rather stringent, which in turn forced the teams to work with a relatively small subset (10% and 25%, respectively) of the total number of protein coding genes. In light of these observations, it is possible that these experiments have undersampled the universe of possibilities and thus may not be entirely representative.

Finally, the work presented in German *et al.*[71] shows an interesting use of high-throughput sequencing in determining the targets of a given microRNA in plants. By sequencing the polyadenylated products of microRNA-driven mRNA decay, the authors are able to recover the location of the cleavage site and thus by extension the identity of the microRNA target. The presented experiments validated several previously predicted targets and identified several novel ones; the latter were confirmed through additional experiments. The methodology is interesting but may prove to be of limited use in animals where microRNA-initiated cleavage of the target's mRNA appears to be less frequent than translational inhibition.

10.4 Estimating Peformance

As an alternative to potentially time-consuming experimental validation using reporter assays, mutation studies, knockout experiments etc., practitioners have employed a variety of computational techniques to determine the performance of the various microRNA target prediction tools. The task is further complicated by the absence of any well-established and widely-accepted protocol for this purpose. Additionally, one can argue that estimating performance with the help of computation is a somewhat ill-defined task for the simple reason that the concept of a 'true positive' is itself not necessarily well defined. For example, if one uses microarray experiments to measure the impact of a microRNA on its targets and to identify these targets, one makes the implicit assumption that microRNA activity is primarily through mRNA degradation. As we have discussed, this may be accurate for plant microRNAs but not representative of animal microRNAs. Indeed, there are many examples in the literature where the microRNA appears to be acting by inhibiting translation and not by degrading the mRNA.[35,37,39,61] Proteomics schemes can of course address this problem but their sensitivity is relatively lower, by comparison, and they capture only targets whose expression levels change substantially.[41,42] No matter whether microarrays or proteomics approaches are used there remains the possibility that the genes whose mRNA or protein levels are lower in the presence of a microRNA are not themselves direct targets of that microRNA. Another way to define a 'true positive' is by relying on a model, e.g. the seed-based one. In this case, one makes the implicit assumption that the model accurately captures reality. In the particular case of the seed, several reports have offered evidence that the seed represents an incomplete model and thus should be used with caution when attempting to evaluate performance.[39,41,42,57,66] Generally, computational estimation of false positives has been estimated using some kind of permutation test on the input features. Occasionally an *additional* assumption is made, namely that genes containing the reverse complement of the microRNA seed are *the true positives*; in this case, false negative rates can also be estimated. In what follows, we briefly outline the choices that were made in the context of some of the developed algorithms.

In TargetScan,[9] the authors used shuffled versions of known microRNA sequences to estimate the false positive rate of their predictions. The shuffled sequences were designed to closely match the following properties of authentic miRNAs: (a) the expected frequency of seed matches in 3' UTRs; (b) the expected frequency of matching to the 3' end of the miRNA; (c) the observed count of seed matches in 3' UTRs; and (d) the predicted free energy of the seed region of the heteroduplex. The reported false positive rate is 31%. The authors also tested 15 predictions using reporter assays and observed suppression for 11 of them. DIANA-microT also used microRNA sequence shuffling but employed a different strategy for estimating false positive rates.[58] For each of the 10 microRNAs that the authors used in their analysis and for which they made predictions, they also generated several shuffled versions of the microRNAs' mature sequences and predicted targets for them. This allowed them to estimate the false positive rate at 39%. MiRanda[19] and PicTar[59] also used shuffled versions of microRNA mature sequences to estimate the false positive rate at 35% and 43%, respectively. In the case of PicTar, the false positive rate is claimed to decrease if a requirement for multiple binding is imposed on the predictions.

Unlike these methods which shuffled the sequences of the targeting microRNA, rna22 instead used shuffled versions of 3' UTR sequences.[30] Care was taken to shuffle only those 3' UTRs that contained a predicted target site that the authors were able to validate using reporter assays: since the 3' UTR contained a validated target site, shuffling it would destroy this target and any target sites that would be reported in it would represent an upper bound for the rate of false positives. In order to account for varying 3' UTR lengths, the authors reported this rate as *number of spurious sites per unit length*: the analysis estimated rna22's false positive rate to be 1 spurious target site every 10,000 nucleotides. In parallel, the authors used a battery of luciferase assays (226 predicted target sites × 12 repetitions) and considered a result to be positive if luciferase suppression averaged 30% or more: of the 226 tested targets 168 met this threshold leading to an alternative estimate of error of $1-168/226 = 26.6\%$.

In GenMiR++, a series of permutation tests was carried out on the miRNA and mRNA labels — the method builds a model of target gene mRNA expression as a function of microRNA expression — and determined the false detection rate to be approximately 2.5%.[60] On the other hand, for NBmiRTar the authors used both a set of known positive and negative examples, and with the help of cross-validation reported 93% sensitivity at 70% specificity, i.e. 30% false positive rate.[64]

Finally, RNAhybrid used an extreme value distribution to model minimum free energies which they used as a proxy for the existence of a microRNA target site.[29] After estimating the distribution's parameters the authors derived an analytical expression for computing the p-values and E-value of predicted target sites.

10.5 Availability of Tools and Predictions

We conclude the presentation with a brief summary of what the various algorithm designers have made available in terms of standalone codes, interactive interfaces or downloadable predictions. Relevant URLs are listed in Table 10.1.

The TargetScan, miRanda, DIANA-microT and mirWIP websites allow the user to search for predicted binding sites for a specific microRNA, gene or microRNA:gene combination. PicTar allows for searching targets of a given miRNA in several organisms. Additionally, batch downloads of pre-compiled target predictions for several genomes are also available for TargetScan and miRanda. The DIANA-microT website also makes available a curated database of experimentally validated targets called TarBase[72] but it appears to be incomplete; for example, the 168 validated targets reported by Miranda *et al.*[30] are not included, as of this writing in spring 2009.

Rna22's website allows the user to explore possible interactions between a microRNA and a putative target sequence by modifying the values of the various parameters. The server also allows the unsupervised batch submission over the web, through a downloadable Java-based utility, of many microRNAs and many target sequences that can be explored in an all-against-all fashion. Additionally, precompiled predictions for several model organisms are available to download.

Table 10.1. Summary of the most representative methods and of their properties. See Section 10.5 for more details.

Reference/Website (if applicable)	Method name	Sequence features					Other features				Validation		
		Perfect seed complementarity	Binding energy	Target site conservation	Multiple target sites	Set of known duplexes	miRNA expression	Gene expression	Set of target genes	Input from experiments	Novel target discovery	FP estimation	FN estimation
24			x	x							6	x	
63	MiTarget		x			x						x	
60 http://www.psi.toronto.edu/genmir/	GenMiR++						x	x	x			x	
65										x			
19 http://microrna.sanger.ac.uk/	miRanda	x	x	x									
30 http://cbcsrv.watson.ibm.com/rna22.html	rna22		x								168/226	x	
58 http://diana.cslab.ece.ntua.gr/microT/	DIANA-microT		x			x				x		x	

Ref	Tool	URL							
47			x	x					x
9	TargetScan	http://www.targetscan.org/	x	x	11/15			x	x
29	RNAhybrid	http://bibiserv.techfak.uni-bielefeld.de/rnahybrid/		x				x	
73	MovingTargets				3/3	x		x	x
34	TargetBoost						x	x	
23	SBM	http://www2.cmp.uea.ac.uk/~jtk/stackbm/				x		x	
66						x		x	x
59	PicTar	http://pictar.mdc-berlin.de/			7/13	x		x	x
62	mirWIP	http://mirtargets.org/				x			
64	NBmiRTar			x			x	x	
67						x	x		
71							x		

From the websites for GenMiR++, RNAhybrid and SBM one can download the codes (sources or binaries) for the corresponding tool. GenMiR++'s website also makes available for download a list of TargetScan predictions that have been ranked and classified by the algorithm whereas the RNAhybrid website allows the interactive submission and exploration of a mature microRNA sequence and a candidate target sequence for possible target sites.

10.6 Conclusion

We have reviewed and discussed a number of microRNA target prediction tools that have been proposed in recent years. Our presentation was confined, by necessity, to the description of the most representative ones within each thematic category and it is by no means exhaustive. As evidenced by the results so far, no proposed solution stands above the rest. Arguably, the problem remains open for all practical intents and purposes.

From a practical standpoint, researchers interested in identifying microRNA targets that they can then pursue experimentally should not confine themselves in using the predictions of only one tool, or, of multiple tools that are based on the same underlying principles; instead, predictions from tools based on *different* premises should be used. Anecdotal evidence has been suggesting that such an approach is rather promising. Nonetheless, it comes with its own complication: it will confidently identify those microRNA targets on which very diverse tools agree but will do so at the likely expense of reduced sensitivity. One possible solution would perhaps be to form the union of predictions made by different tools and then experimentally work with a sub-sample of this union that does not include the intersection of the various sets.

In closing, we hope that we have succeeded in conveying a good summary view of the various attempts to tackle this problem. Without a doubt, a lot of progress has been made in the last few years, but a lot more needs to be done. It is our belief that the years ahead will witness great achievements by experimental biologists studying microRNA:mRNA interactions. These should in turn inform computational approaches by providing invaluable data points on which to train their methods and eventually improve the methods' performance.

References

1. Bartel D.P. (2009). MicroRNAs: Target recognition and regulatory functions. *Cell* **136**, 215–233.
2. Lee R.C., Feinbaum R.L. and Ambros V. (1993). The *C. elegans* heterochronic gene *lin-4* encodes small RNAs with antisense complementarity to lin-14. *Cell* **75**, 843–854.
3. Wightman B., Ha I. and Ruvkun G. (1993). Posttranscriptional regulation of the heterochronic gene lin-14 by *lin-4* mediates temporal pattern formation in *C. elegans*. *Cell* **75**, 855–862.
4. Reinhart B.J., Slack F.J., Basson M. *et al.* (2000). The 21-nucleotide let-7 RNA regulates developmental timing in *Caenorhabditis elegans*. *Nature* **403**, 901–906.
5. Reinhart B.J., Weinstein E.G., Rhoades M.W. *et al.* (2002). MicroRNAs in plants. *Genes Dev* **16**, 1616–1626.
6. Lagos-Quintana M., Rauhut R., Meyer J. *et al.* (2003). New microRNAs from mouse and human. *RNA* **9**, 175–179.
7. Pfeffer S., Sewer A., Lagos-Quintana M. *et al.* (2005). Identification of microRNAs of the herpesvirus family. *Nat Methods* **2**, 269–276.
8. Pfeffer S., Zavolan M., Grasser F.A. (2004). Identification of virus-encoded microRNAs. *Science* **304**, 734–736.
9. Lewis B.P., Shih I.H., Jones-Rhoades M.W. *et al.* (2003). Prediction of mammalian microRNA targets. *Cell* **115**, 787–798.
10. Rigoutsos I. (2009). New tricks for animal microRNAS: Targeting of amino acid coding regions at conserved and nonconserved sites. *Cancer Res* **69**, 3245–3248.
11. Wang Y., Juranek S., Li H. *et al.* (2008). Structure of an argonaute silencing complex with a seed-containing guide DNA and target RNA duplex. *Nature* **456**, 921–926.
12. Ha I., Wightman B. and Ruvkun G. (1996). A bulged *lin-4*/lin-14 RNA duplex is sufficient for *Caenorhabditis elegans* lin-14 temporal gradient formation. *Genes Dev* **10**, 3041–3050.
13. Vella M.C., Choi E.Y., Lin S.Y. *et al.* (2004). The *C. elegans* microRNA let-7 binds to imperfect let-7 complementary sites from the *lin-41* 3'UTR. *Genes Dev* **18**, 132–137.
14. Bartel D.P. (2004). MicroRNAs: Genomics, biogenesis, mechanism, and function. *Cell* **116**, 281–297.
15. Grimson A., Farh K.K., Johnston W.K. *et al.* (2007). MicroRNA targeting specificity in mammals: Determinants beyond seed pairing. *Mol Cell* **27**, 91–105.

16. Brennecke J., Stark A., Russell R.B. *et al.* (2005). Principles of microRNA-target recognition. *PLoS Biol* 3, e85.

17. Hofacker I.L., Fontana W., Stadler P. *et al.* (1994). Fast folding and comparison of RNA secondary structures. *Monatshefte f Chemie* 125, 167–188.

18. Burgler C. and Macdonald P.M. (2005). Prediction and verification of microRNA targets by MovingTargets, a highly adaptable prediction method. *BMC Genomics* 6, 88.

19. Enright A.J., John B. and Gaul U. (2003). MicroRNA targets in Drosophila. *Genome Biol* 5, R1.

20. Smith T.F. and Waterman M.S. (1981). Identification of common molecular subsequences. *J Mol Biol* 147, 195–197.

21. Needleman S.B. and Wunsch C.D. (1970). A general method applicable to the search for similarities in the amino acid sequence of two proteins. *J Mol Biol* 48, 443–453.

22. John B., Enright A.J., Aravin A. *et al.* (2004). Human MicroRNA targets. *PLoS Biol* 2, e363.

23. Moxon S., Moulton V. and Kim J. (2008). A scoring matrix approach to detecting miRNA target sites. *Algorithm Mol Biol* 3, 3.

24. Stark A., Brennecke J., Russell R.B. *et al.* (2003). Identification of Drosophila MicroRNA targets. *PLoS Biol* 1, E60.

25. Eddy S.R. (1996). Hidden Markov models. *Curr Opin Struct Biol* 6, 361–365.

26. Mathews D.H., Sabina J., Zuker M. *et al.* (1999). Expanded sequence dependence of thermodynamic parameters improves prediction of RNA secondary structure. *J Mol Biol* 288, 911–940.

27. Zuker M. (2003). Mfold web server for nucleic acid folding and hybridization prediction. *Nucleic Acids Res* 31, 3406–3415.

28. Rajewsky N. and Socci N.D. (2004). Computational identification of microRNA targets. *Dev Biol* 267, 529–535.

29. Rehmsmeier M., Steffen P., Hochsmann M. *et al.* (2004). Fast and effective prediction of microRNA/target duplexes. *RNA* 10, 1507–1517.

30. Miranda K.C., Huynh T., Tay Y. *et al.* (2006). A pattern-based method for the identification of MicroRNA binding sites and their corresponding heteroduplexes. *Cell* 126, 1203–1217.

31. Griffiths-Jones S. (2006). miRBase: The microRNA sequence database. *Methods Mol Biol* 342, 129–138.

32. Rigoutsos I., Floratos A., Parida L. *et al.* (2000). The emergence of pattern discovery techniques in computational biology. *Metab Eng* 2, 159–177.

33. Rigoutsos I. and Floratos A. (1998). Combinatorial pattern discovery in biological sequences: The TEIRESIAS algorithm. *Bioinformatics* 14, 55–67.
34. Saetrom O., Snove O., Jr. and Saetrom P. (2005). Weighted sequence motifs as an improved seeding step in microRNA target prediction algorithms. *RNA* 11, 995–1003.
35. Duursma A.M., Kedde M., Schrier M. *et al.* (2008). MiR-148 targets human DNMT3b protein coding region. *RNA* 14, 872–877.
36. Forman J.J., Legesse-Miller A. and Coller H.A. (2008). A search for conserved sequences in coding regions reveals that the let-7 microRNA targets Dicer within its coding sequence. *Proc Natl Acad Sci USA* 105, 14879–14884.
37. Lal A., Kim H.H., Abdelmohsen K. *et al.* (2008). p16(INK4a) translation suppressed by miR-24. *PLoS ONE* 3, e1864.
38. Shen W.F., Hu Y.L., Uttarwar L. *et al.* (2008). MicroRNA-126 regulates HOXA9 by binding to the homeobox. *Mol Cell Biol* 28, 4609–4619.
39. Tay Y., Zhang J., Thomson A.M. *et al.* (2008). MicroRNAs to Nanog, Oct4 and Sox2 coding regions modulate embryonic stem cell differentiation. *Nature* 455, 1124–1128.
40. Tay Y.M., Tam W.L., Ang Y.S. *et al.* (2008). MicroRNA-134 modulates the differentiation of mouse embryonic stem cells, where it causes post-transcriptional attenuation of Nanog and LRH1. *Stem Cells* 26, 17–29.
41. Baek D., Villen J., Shin C. *et al.* (2008). The impact of microRNAs on protein output. *Nature* 455, 64–71.
42. Selbach M., Schwanhausser B. and Thierfelder N. (2008). Widespread changes in protein synthesis induced by microRNAs. *Nature* 455, 58–63.
43. Doolittle R.F., Hunkapiller M.W., Hood L.E. *et al.* (1983). Simian sarcoma virus onc gene, v-sis, is derived from the gene (or genes) encoding a platelet-derived growth factor. *Science* 221, 275–277.
44. Gusfield D. (1997). *Algorithms on Strings, Trees, and Sequences: Computer Science and Computational Biology.* Cambridge University Press, Cambridge and New York.
45. Mount D.W. (2004). *Bioinformatics: Sequence and Genome Analysis.* [2nd edition.] Cold Spring Harbor Laboratory Press, Cold Spring Harbor, New York.
46. Xie X., Lu J., Kulbokas E.J. *et al.* (2005). Systematic discovery of regulatory motifs in human promoters and 3′ UTRs by comparison of several mammals. *Nature* 434, 338–345.
47. Kheradpour P., Stark A., Roy S. *et al.* (2007). Reliable prediction of regulator targets using 12 Drosophila genomes. *Genome Res* 17, 1919–1931.

48. ENCODE (2007). Identification and analysis of functional elements in 1% of the human genome by the ENCODE pilot project. *Nature* **447**, 799–816.

49. Fisher S., Grice E.A. and Vinton R.M. (2006). Conservation of RET regulatory function from human to zebrafish without sequence similarity. *Science* **312**, 276–279.

50. McGaughey D.M., Vinton R.M., Huynh J. *et al.* (2008). Metrics of sequence constraint overlook regulatory sequences in an exhaustive analysis at phox2b. *Genome Res* **18**, 252–260.

51. Rigoutsos I., Huynh T., Miranda K. *et al.* (2006). Short blocks from the non-coding parts of the human genome have instances within nearly all known genes and relate to biological processes. *Proc Natl Acad Sci USA* **103**, 6605–6610.

52. Tsirigos A. and Rigoutsos I. (2008). Human and mouse introns are linked to the same processes and functions through each genome's most frequent non-conserved motifs. *Nucleic Acids Res* **36**, 3484–3493.

53. Doench J.G. and Sharp P.A. (2004). Specificity of microRNA target selection in translational repression. *Genes Dev* **18**, 504–511.

54. Kertesz M., Iovino N., Unnerstall U. *et al.* (2007). The role of site accessibility in microRNA target recognition. *Nat Genet* **39**, 1278–1284.

55. Robins H., Li Y. and Padgett R.W. (2005). Incorporating structure to predict microRNA targets. *Proc Natl Acad Sci USA* **102**, 4006–4009.

56. Zhao Y., Samal E. and Srivastava D. (2005). Serum response factor regulates a muscle-specific microRNA that targets Hand2 during cardiogenesis. *Nature* **436**, 214–220.

57. Didiano D. and Hobert O. (2006). Perfect seed pairing is not a generally reliable predictor for miRNA-target interactions. *Nat Struct Mol Biol* **13**, 849–851.

58. Kiriakidou M., Nelson P.T., Kouranov A. *et al.* (2004). A combined computational-experimental approach predicts human microRNA targets. *Genes Dev* **18**, 1165–1178.

59. Krek A., Grun D., Poy M.N. *et al.* (2005). Combinatorial microRNA target predictions. *Nat Genet* **37**, 495–500.

60. Huang J.C., Morris Q.D. and Frey B.J. (2007). Bayesian inference of MicroRNA targets from sequence and expression data. *J Comput Biol* **14**, 550–563.

61. Filipowicz W., Bhattacharyya S.N. and Sonenberg N. (2008). Mechanisms of post-transcriptional regulation by microRNAs: Are the answers in sight? *Nat Rev Genet* **9**, 102–114.

62. Hammell M., Long D., Zhang L. *et al.* (2008). mirWIP: MicroRNA target prediction based on microRNA-containing ribonucleoprotein-enriched transcripts. *Nat Methods* **5**, 813–819.

63. Kim S.K., Nam J.W., Rhee J.K. *et al.* (2006). miTarget: MicroRNA target gene prediction using a support vector machine. *BMC Bioinformatics* 7, 411.
64. Yousef M., Jung S., Kossenkov A.V. *et al.* (2007). Naive Bayes for microRNA target predictions — machine learning for microRNA targets. *Bioinformatics* 23, 2987–2992.
65. Vatolin S., Navaratne K. and Weil R.J. (2006). A novel method to detect functional microRNA targets. *J Mol Biol* 358, 983–996.
66. Easow G., Teleman A.A. and Cohen S.M. (2007). Isolation of microRNA targets by miRNP immunopurification. *RNA* 13, 1198–1204.
67. Karginov F.V., Conaco C., Xuan Z. *et al.* (2007). A biochemical approach to identifying microRNA targets. *Proc Natl Acad Sci* 104, 19291–19296.
68. Lim L.P., Lau N.C., Garrett-Engele P. *et al.* (2005). Microarray analysis shows that some microRNAs downregulate large numbers of target mRNAs. *Nature* 433, 769–773.
69. Mann M. (2006). Functional and quantitative proteomics using SILAC. *Nat Rev Mol Cell Biol* 7, 952–958.
70. Ong S.E., Blagoev B., Kratchmarova I. *et al.* (2002). Stable isotope labeling by amino acids in cell culture, SILAC, as a simple and accurate approach to expression proteomics. *Mol Cell Proteomics* 1, 376–386.
71. German M.A., Pillay M., Jeong D.H. *et al.* (2008). Global identification of microRNA-target RNA pairs by parallel analysis of RNA ends. *Nat Biotechnol* 26, 941–946.
72. Papadopoulos G.L., Reczko M., Simossis V.A. *et al.* (2009). The database of experimentally supported targets: A functional update of TarBase. *Nucleic Acids Res* 37, D155–158.
73. Burgler C. and Macdonald P. (2005). Prediction and verification of microRNA targets by MovingTargets, a highly adaptable prediction method. *BMC Genomics* 6, 88.

11

Non-Coding RNAs in Cancer — The Other Part of the Story

Muller Fabbri* and George A. Calin[†,‡]

Despite the leading role of microRNAs (miRNAs) as cancer-related non-coding RNAs (ncRNAs) in published research, recently new categories of not-translated RNAs have emerged. The involvement of the long non-coding ultraconserved genes (UCGs) in cancer is suggested by their frequent location in cancer-associated genomic regions (CAGRs), and their aberrant expression in several human cancers with respect to the normal tissue counterpart. Signatures of deregulated UCGs are cancer-specific and have prognostic implications. Similarly to miRNAs, UCGs can act as oncogenes or tumour suppressor genes, and their expression is under the control of miRNAs. Another group of ncRNAs, the short germline specific PIWI-associated RNAs (piRNAs), involved in the regulation of transposable elements and messenger RNAs (mRNAs), might have implications for human carcinogenesis, although this connection is at the moment still elusive.

* Department of Molecular Virology, Immunology, and Medical Genetics and Comprehensive Cancer Center, Ohio State University, Columbus, OH 43210, USA. †Department of Experimental Therapeutics and Cancer Genetics, University of Texas, M.D. Anderson Cancer Center, Houston, TX 77030, USA. ‡E-mail: gcalin@ mdanderson.org.

11.1 Introduction

Cancer is a very complex genetic disease, characterised by multiple genetic abnormalities in both protein-coding and non-coding oncogenes (OGs), and tumour suppressor genes (TSGs).[1–3] An increasing number of studies prove without doubt that non-coding RNAs (ncRNAs) are among the most aberrantly expressed genes in cancer, with respect to normal tissues. Despite microRNAs (miRNAs) being the most extensively-studied, deregulated ncRNAs in human malignancies,[4–7] other aberrantly expressed ncRNAs have been recently implicated in the genesis of the malignant phenotype. To exemplify this, in this chapter we focus on two very different categories of non-coding transcripts. Firstly, the genomic ultraconserved regions (UCRs) encode a particular set of long ncRNAs (hundreds to thousands of bases transcripts), whose expression is altered in human cancers, named as ultra-conserved genes (UCGs, or transcribed UCRs).[8] UCRs are a subset of genomic sequences which are strictly conserved among orthologous regions of the human, rat and mouse genomes,[9] and due to their non-coding nature they have been considered for a long time as the 'dark matter' of the human genome, being initially considered silent evolution relicts.[10] The development of high-throughput methods able to detect the expression of ncRNAs has shown not only that these regions are transcribed, but also that their expression is different in normal versus neoplastic tissues. The tumour-specificity of this signature harbours several clinical implications, which could lead to the development of new therapeutic strategies. Secondly, among the most recently identified ncRNAs involved in gene regulation, there are the PIWI-associated RNAs (piRNAs), which are 24- to 33-nt long ncRNAs, produced in a Dicer-independent manner,[11–15] whose function is to protect the germline from mobile genome invaders such as transposons.[16–18] Similarly to miRNAs, piRNAs have repressive regulatory functions, but the mechanism of piRNA-based gene silencing remains unknown.[19] The main differential characteristics between miRNAs, UCGs and piRNAs are listed in Table 11.1.

A better understanding of the physiology of UCRs and piRNAs will lead to a better comprehension of their regulatory mechanisms and of their aberrancies in human cancers. This chapter focuses on the studies that have investigated the involvement of these ncRNAs in human cancers,

Table 11.1.   Characteristics of the three main groups of ncRNAs.

Property	miRNAs	UCRs	piRNAs
Length	20–21nt	>200nt	24–33nt
Binding protein	Argonaute family	Unknown	PIWI
Number of distinct sequences	>1000	481	200,000 estimated 50,000 known
Expression patterns	Most cell types	Most cell types	Germline cells
Biogenesis	Dicer dependent	Unknown	Dicer independent
Conservation	High	Very high	Limited
Function	Post-transcriptional regulation of target mRNAs	Gene regulation (?)	Transposon silencing, possible mRNA degradation

and the possible regulatory interactions among them, all of which could open new exciting scenarios in the diagnosis and treatment of human neoplasms.

11.2 UCGs: Location and Expression Profiles Prove a Cancer Link

First discovered by Bejerano *et al.* in 2004, by bioinformatics comparison of the genomes of mouse, rat and human,[9,10] UCRs are involved in promoting the expression of several genes and in regulating genetic alternative splicing.[20] Phylogenetically, UCRs date from a very early period in vertebrate evolution, since no orthologous counterparts have been described in sea squirts, flies or worms.[20] Nevertheless, a distal enhancer and an ultraconserved exon were described from a novel retrotransposon active in lobe-finned fishes and terrestrial vertebrates more than 400 million years ago and still active in the 'living fossil' coelacanth.[20] At a genomic level, UCRs are located in both intra- and intergenic regions,[9] and are classified as nonexonic (N, with no evidence of encoding protein), exonic (E, when they overlap known protein-coding genes), or possibly exonic (P, when there is non-conclusive evidence of overlap with protein-coding genes).[9] Of the 481 UCRs longer than 200 bp described by Bejerano *et al.*, 53% have been classified as N-UCRs, and 47% as E-UCRs or P-UCRs.[9] By using a microarray analysis, Calin *et al.*, found that 93% of the UCGs are transcribed in at least one tissue, and that transcribed UCRs are expressed in normal human tissues both ubiquitously and in a tissue-specific manner.[8] No difference in the percentage of transcribed UCRs was observed among E- (41%), P- (33%) or N-UCRs (30%).[8] The expression of UCGs was also investigated in a panel of 173 samples, including 133 human cancers and 40 corresponding normal tissues.[8] Chronic lymphocytic leukaemia (CLL), colorectal (CRC) and hepatocellular (HCC) carcinomas were compared to their normal counterpart, and tissue- and cancer-specific signatures of differentially expressed UCGs were observed.[8] In particular, CLL signature consisted of 19 differentially expressed UCGs (8 up- and 11 downregulated), CRC signature of 61 UCGs (59 up- and 2 downregulated), and HCC signature of 8 (3 up- and 5 downregulated).[8] Similarly to what was described for miRNAs,[21] UCG expression profiles are cancer specific and might be used to differentiate human tumours.

At a genomic level, UCRs are more frequently located in cancer associated genomic regions (CAGRs),[8] which include fragile sites (FRAs), minimal regions of amplification, loss of heterozygosity (LOH), common breakpoint regions in or near OGs or TSGs.[22] This distribution is comparable to that reported for miRNAs,[23] since both groups of ncRNAs significantly correlate with the distribution of human papilloma virus 16 (HPV16) integration sites (which frequently occurs in FRAs). Finally, UCGs belonging to a specific cancer signature are frequently located in CAGRs specifically associated to that type of cancer. For instance, uc.349A (A = transcribed from the antisense strand), and uc.352 are differentially expressed in normal versus malignant B-CLL CD5-positive cells,[8] and they map at 13q21.33-q22.2, a chromosomal region linked to susceptibility to familial CLL,[24] but with no identified mutations in any of the 13 protein-coding genes located in this interval. Overall, these data indicate that UCRs are differentially expressed in human normal and neoplastic tissues and harbour some characteristics of cancer susceptibility genes.

11.3 UCGs as Oncogenes or Tumour Suppressor Genes

Cancer-associated signatures of UCGs consist of up- and downregulated transcripts (Table 11.2) suggesting a possible role as OGs and TSGs, respectively (Figure 11.1). Specifically, the oncogenic nature of uc.73A was demonstrated in a CRC model, since this ncRNA is one of the most statistically significant upregulated UCG in CRC ($p < 0.001$).[8] COLO-320 CRC

Table 11.2.   Most significantly differentially expressed UCRs in Colorectal Cancer (CRC) and Chronic Lymphocytic Leukemia (CLL).

UCR name	Type	CRC	CLL	CAGR location
Uc.29	N	H	Normal	No
Uc.73	P	H	L	No
Uc.111	P	H	Normal	Yes
Uc.112	N	H	Normal	No
Uc.134	P	H	Normal	No
Uc.135	E	Normal	L	Yes
Uc.206	N	H	Normal	No
Uc.230	P	H	Normal	No
Uc.233	E	Normal	L	No
Uc.291	P	Normal	L	No
Uc.292	E	H	Normal	No
Uc.339	P	H	Normal	Yes
Uc.341	E	H	Normal	Yes
Uc.388	N	H	Normal	No
Uc.399	N	H	Normal	No
Uc.420	E	H	Normal	No

N = Nonexonic, P = possibly exonic, E = exonic, H = high versus normal counterpart, L = low versus normal counterpart, CAGR = Cancer Associated Genomic Region.

cell line expresses high levels of uc.73A versus normal colon tissue, whereas SW620 CRC cells express uc.73A levels comparable to that of normal colon cells. By silencing uc.73A expression with small interfering RNAs, Calin *et al.* observed a reduced growth of COLO-320 cells and an increase of sub-G1 fraction of cells with increased apoptosis with respect to COLO-320 cells not treated. No effect was observed in SW620 cells, suggesting that uc.73A may act as an OG in CRC by increasing the number of malignant cells as a consequence of reduced apoptosis. Although no functional studies are yet available, one can hypothesise a function for those UCRs downregulated in cancer as TSGs (Figure 11.1).

11.4 Regulation of UCG Expression

The regulatory mechanisms underlying UCR expression are still poorly understood. Nevertheless, an intriguing interaction between UCRs and

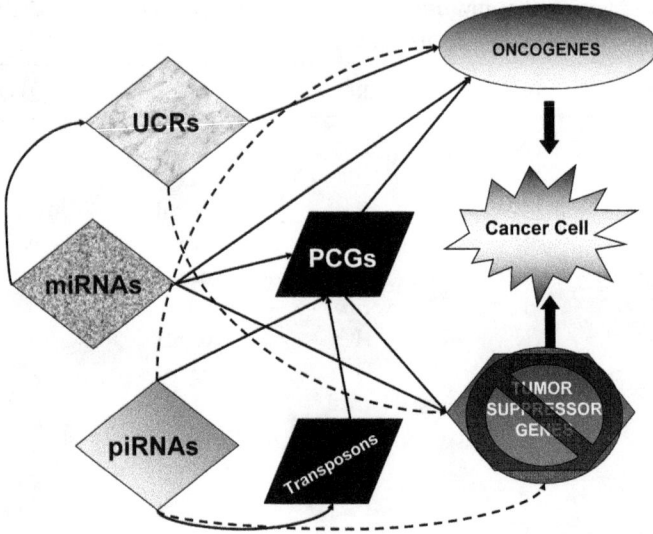

Figure 11.1. Involvement of ncRNAs in human cancerogenesis. In cancer cells, oncogenes (OGs) are predominant over the action of tumour suppressor genes (TSGs). Protein-coding genes (PCGs) can function as OGs or TSGs. Non-coding RNAs (in diamonds) involved in human tumouriogenesis are miRNAs, UCRs (ultraconserved regions) and, we postulate, the possible roles of piRNAs (PIWI-associated RNAs). MiRNAs regulate the expression of PCGs and can act either as OGs or as TSGs or both. UCRs, regulated by miRNAs, can act as OGs, whereas their role as TSGs can be hypothesised but has not yet been experimentally proven. PiRNAs regulate the function of transposons, which, in turn, can affect PCG expression, but they can also directly regulate PCG expression. A role for piRNAs as OGs or TSGs has not been clearly demonstrated yet.
Source: Solid Arrow = experimentally proven interaction/function; Dashed Arrow = not proven interaction/function.

miRNAs has been demonstrated, suggesting the existence of a network of ncRNAs with reciprocal controlling functions (Figure 11.1). Similarly to what was described for miRNAs,[25] a specific signature composed of 5 UCRs (namely uc.269A, uc.160, uc.215, uc.346A and uc.348) differentiates between CLLs with worse (high levels of Zap-70) and better (low levels of Zap-70) prognosis.[8] A negative correlation was observed between these 5 UCGs and the miRNA expression signature previously reported in CLL. This finding prompted to check whether complementary matching sites for miRNAs were located in these UCG sequences. By sequence alignments, they found that 3 out of 5 UCGs had significant antisense complementarity with 5 out of 13 miRNAs from the signature. The authors showed that miR-155 (frequently upregulated in several human malignancies,[26,27] including CLL[8]), directly targets uc.346A and uc.160, while miR-24-1 directly targets uc.160 both *in vitro* and in CLL patients.[8] This interaction between ncRNAs adds a further level of complexity in the regulation of ncRNA expression, and may have a biological and clinical significance for cancer patients.

11.5 Sequence Variations in UCRs and Cancer Risk

As for miRNAs, it has been demonstrated that single nucleotide polymorphisms (SNPs) affect their levels of expression and predispose them to several kinds of tumours;[28] an analysis of SNPs in UCR and cancer risk has been already published.[29] SNPs are strongly under-represented within UCRs, since only 6 SNPs have been validated in 481 examined UCRs.[9] By comparing BRCA1/2 mutation-negative index patients from 1,214 German breast cancer families and 2,084 German female blood donors, Yang *et al.*, showed that 2 SNPs — the rs2056116 (G versus A) located in the region of uc.140, and the rs9572903 (G versus A) in the region of uc.353 — were associated with breast cancer risk. While rs9573903 showed a borderline association with breast cancer risk, the association of rs2056116 was stronger.[29] In particular, when all cases and controls were subdivided according to the threshold age of 50 years, an increased risk of breast cancer was observed in carriers of the G allele under 50 years of age.[29] No significant risk association was observed in the ≥50 years' group.[29] Interestingly, Catucci *et al.* did not support the association of these 2 SNPs and breast cancer risk in a population of 737 Italian female familial breast cancer cases negative for mutations in BRCA1/2 genes compared to 1,245 Italian female blood donors.[30] The conflicting results could be explained by differences in the allele frequencies between the German and the Italian control populations (0.39 vs 0.34, respectively, $p = 0.00007$) and by statistical fluctuations. Another limit of both studies is the number of patients involved in the analysis. Large multiple-centre studies are warranted to evaluate the effect of SNPs on breast cancer risk. Moreover, no study has yet addressed whether these SNPs affect the expression of transcribed UCRs. A better understanding of how SNPs impact on the expression of UCRs might provide a deeper insight on the regulatory mechanisms of these ncRNAs and could have prognostic and therapeutic implications for cancer patients.

11.6 PiRNAs and Cancer

In 2006, a new class of germline-specific small ncRNAs (24 to 33 nucleotides long), named PIWI-interacting RNAs (piRNAs) was discovered and found to bind the testis-specific PIWI protein orthologues from mouse and rat[11–15] and in being abnormally expressed in human cancers. Differently from miRNAs and UCRs, a clear, direct involvement of piRNAs in human carcinogenesis has not yet been demonstrated. Nevertheless, numerous genetic studies point towards a role of Piwi family members in controlling transposons[31–33] (Figure 11.1). In mammals, transposons constitute up to 50% of the genome.[34] When mobilised, they can disrupt protein-coding genes, cause chromosomal breakage, large-scale genomic rearrangement and modify transcriptional regulatory networks.[35] Approximately 4% of human genes possess some transposon-derived coding sequences,[36] revealing the potentiality of these transposable elements to generate genetic divergences that could drive evolution. Moreover, transposons contain their own transcriptional regulatory elements[37] and their motility can impact on the expression patterns of neighbouring genes[38] (Figure 11.1). Recently, Aravin *et al.*, have identified a piRNA pathway involved as a specific determinant of DNA methylation in germ cells,[39] suggesting an intriguing connection between piRNAs and epigenesis. Finally, Piwi-class Argonaute-piRNA complexes, similarly to siRNA or miRNA-Argonaute complexes, can cleave perfectly-matched RNA targets *in vitro*,[15,17,40] and might also trigger the translational silencing of imperfectly-matched targets, including mRNAs from single-copy genes, as shown in Drosophila for the gene *oskar*, involved in embryonic patterning.[41] Overall, the control of transposon motility achieved both by means of a direct basepair interaction, and by epigenetic mechanisms, and the ability to trigger mRNA silencing, are suggestive of a possible role of piRNAs in carcinogenesis (Figure 11.1).

Intriguingly, the human homologue of the subgroup of the Argonaute proteins involved in piRNA processing, named *HIWI* (the human Piwi gene that encodes a protein responsible for stem cell self-renewal) is precisely located in the genomic region 12q24.33 that is genetically linked with the development of testicular germ cell tumours of adolescents and adults.

In addition, *HIWI* is overexpressed in the majority of testicular seminomas analysed to date,[42] and high-level expression of Hiwi mRNA identifies soft tissue sarcoma patients at high risk of tumour-related death.[43] Furthermore, alterations in mRNA expression of Hiwi can increase the risk of tumour-related death in male ductal adenocarcinoma of the pancreas patients.[44] Expression of *HIWI* in human gastric cancer was associated with proliferating cells.[45] These correlations raise the possibility that piRNAs could represent another example of cancer-altered small ncRNAs, specifically in relation to PIWI overexpression. Although at the moment in which this chapter is being written, no evidence of a direct piRNA-cancer connection has been provided, the rationale to design experiments aimed at clarifying the role of these ncRNAs in cancer is compelling and more studies in this direction are warranted.

11.7 Conclusion

In recent years, the emerging roles of ncRNAs in human tumours has challenged the central dogma of molecular oncology, according to which cancer is a genetic disease ultimately resulting from an imbalance between protein-coding oncogenes or tumour suppressor genes. The miRNA revolution has indeed introduced a further level of complexity in post-transcriptional gene regulation, widening our comprehension of human malignancies. The recent discovery that other ncRNAs (such as UCGs and piRNAs) might play an important role in controlling gene expression, and that intertwined regulatory networks exist among different categories of ncRNAs (such as miRNAs and UCRs), represent important improvements in the effort to decode the complex laws of cancer. Even so, several issues still need to be addressed before these ncRNAs can be used as prognostic markers and as anti-cancer drugs. The progressively increasing understanding of the implications of ncRNAs for the malignant phenotype represents the essential background to achieve the goal for a better treatment of cancer patients.

Acknowledgments

G.A.C. is supported as a fellow at the University of Texas M.D. Anderson Research Trust; as a fellow of the University of Texas System Regents Research Scholar; by the Ladjevardian Regents Research Scholar Fund; by an NIH grant 1R01CA135444; and by 2009 Seena Magowitz — Pancreatic Cancer Action Network — AACR Pilot Grant. Dr. Fabbri is supported by a 2009 Kimmel Scholar Award.

References

1. Calin G.A. and Croce C.M. (2006). Genomics of chronic lymphocytic leukemia microRNAs as new players with clinical significance. *Semin Oncol* **33**, 167–173.
2. Croce C.M. (2008). Oncogenes and cancer. *N Engl J Med* **358**, 502–511.
3. Esquela-Kerscher A. and Slack F.J. (2006). Oncomirs — MicroRNAs with a role in cancer. *Nat Rev Cancer* **6**, 259–269.
4. Ambros V. (2004). The functions of animal microRNAs. *Nature* **431**, 350–355.
5. Fabbri M., Croce C.M. and Calin G.A. (2008). MicroRNAs. *Cancer J* **14**, 1–6.
6. Fabbri M., Garzon R., Andreeff M. *et al.* (2008). MicroRNAs and noncoding RNAs in hematological malignancies: Molecular, clinical and therapeutic implications. *Leukemia* **22**, 1095–1105.
7. Slack F.J. and Weidhaas J.B. (2008). MicroRNA in cancer prognosis. *N Engl J Med* **359**, 2720–2722.
8. Calin G.A., Liu C.G., Ferracin M. *et al.* (2007). Ultraconserved regions encoding ncRNAs are altered in human leukemias and carcinomas. *Cancer Cell* **12**, 215–229.
9. Bejerano G., Pheasant M., Makunin I. *et al.* (2004). Ultraconserved elements in the human genome. *Science* **304**, 1321–1325.
10. Bejerano G., Haussler D. and Blanchette M. (2004). Into the heart of darkness: Large-scale clustering of human non-coding DNA. *Bioinformatics* **20**, 140–148.
11. Aravin A., Gaidatzis D., Pfeffer S. *et al.* (2006). A novel class of small RNAs bind to MILI protein in mouse testes. *Nature* **442**, 203–207.
12. Girard A., Sachidanandam R., Hannon G.J. *et al.* (2006). A germline-specific class of small RNAs binds mammalian Piwi proteins. *Nature* **442**, 199–202.
13. Grivna S.T., Beyret E., Wang Z. *et al.* (2006). A novel class of small RNAs in mouse spermatogenic cells. *Genes Dev* **20**, 1709–1714.
14. Grivna S.T., Pyhtila B. and Lin H. (2006). MIWI associates with translational machinery and PIWI-interacting RNAs (piRNAs) in regulating spermatogenesis. *Proc Natl Acad Sci USA* **103**, 13415–13420.

15. Saito K., Nishida K.M., Mori T. *et al.* (2006). Specific association of Piwi with rasiRNAs derived from retrotransposon and heterochromatic regions in the Drosophila genome. *Genes Dev* **20**, 2214–2222.

16. Watanabe T., Takeda A., Tsukiyama T. *et al.* (2006). Identification and characterization of two novel classes of small RNAs in the mouse germline: Retrotransposon-derived siRNAs in oocytes and germline small RNAs in testes. *Genes Dev* **20**, 1732–1743.

17. Lau N.C., Seto A.G., Kim J. *et al.* (2006). Characterization of the piRNA complex from rat testes. *Science* **313**, 363–367.

18. O'Donnell K.A. and Boeke J.D. (2007). Mighty Piwis defend the germline against genome intruders. *Cell* **129**, 37–44.

19. Klattenhoff C. and Theurkauf W. (2008). Biogenesis and germline functions of piRNAs. *Development* **135**, 3–9.

20. Bejerano G., Lowe C.B., Ahituv N. *et al.* (2006). A distal enhancer and an ultraconserved exon are derived from a novel retroposon. *Nature* **441**, 87–90.

21. Liu C.G., Calin G.A., Meloon B. *et al.* (2004). An oligonucleotide microchip for genome-wide microRNA profiling in human and mouse tissues. *Proc Natl Acad Sci USA* **101**, 9740–9744.

22. Rossi S., Sevignani C., Nnadi S.C. *et al.* (2008). Cancer-associated genomic regions (CAGRs) and noncoding RNAs: Bioinformatics and therapeutic implications. *Mamm Genome* **19**, 526–540.

23. Calin G.A., Sevignani C., Dumitru C.D. *et al.* (2004). Human microRNA genes are frequently located at fragile sites and genomic regions involved in cancers. *Proc Natl Acad Sci USA* **101**, 2999–3004.

24. Ng D., Toure O., Wei M.H. *et al.* (2007). Identification of a novel chromosome region, 13q21.33-q22.2, for susceptibility genes in familial chronic lymphocytic leukemia. *Blood* **109**, 916–925.

25. Calin G.A., Ferracin M., Cimmino A. *et al.* (2005). A microRNA signature associated with prognosis and progression in chronic lymphocytic leukemia. *N Engl J Med* **353**, 1793–1801.

26. Volinia S., Calin G.A., Liu C.G. *et al.* (2006). A microRNA expression signature of human solid tumors defines cancer gene targets. *Proc Natl Acad Sci USA* **103**, 2257–2261.

27. Yanaihara N., Caplen N., Bowman E. *et al.* (2006). MicroRNA signature in lung cancer diagnosis and prognosis. *Cancer Cell* **9**, 189–198.

28. Fabbri M., Valeri N. and Calin G.A. (2009). MicroRNAs and genomic variations: From Proteus tricks to Prometheus gift. *Carcinogenesis* **30**, 912–917.

29. Yang R., Frank B., Hemminki K. *et al.* (2008). SNPs in ultraconserved elements and familial breast cancer risk. *Carcinogenesis* **29**, 351–355.
30. Catucci I., Verderio P., Pizzamiglio S. *et al.* (2009). SNPs in ultraconserved elements and familial breast cancer risk. *Carcinogenesis* **30**, 544–545.
31. Sarot E., Payen-Groshene G., Bucheton A. *et al.* (2004). Evidence for a piwi-dependent RNA silencing of the gypsy endogenous retrovirus by the *Drosophila melanogaster* flamenco gene. *Genetics* **166**, 1313–1321.
32. Savitsky M., Kwon D., Georgiev P. *et al.* (2006). Telomere elongation is under the control of the RNAi-based mechanism in the Drosophila germline. *Genes Dev* **20**, 345–354.
33. Reiss D., Josse T., Anxolabehere D. *et al.* (2004). Aubergine mutations in *Drosophila melanogaster* impair P cytotype determination by telomeric P elements inserted in heterochromatin. *Mol Genet Genomics* **272**, 336–343.
34. Kazazian H.H. (2004). Mobile elements: Drivers of genome evolution. *Science* **303**, 1626–1632.
35. McClintock B. (1951). Chromosome organization and genic expression. *Cold Spring Harb Symp Quant Biol* **16**, 13–47.
36. Nekrutenko A. and Li W.H. (2001). Transposable elements are found in a large number of human protein-coding genes. *Trends Genet* **17**, 619–621.
37. White S.E., Habera L.F. and Wessler S.R. (1994). Retrotransposons in the flanking regions of normal plant genes: A role for copia-like elements in the evolution of gene structure and expression. *Proc Natl Acad Sci USA* **91**, 11792–11796.
38. Landry J.R., Medstrand P. and Mager D.L. (2001). Repetitive elements in the 5′ untranslated region of a human zinc-finger gene modulate transcription and translation efficiency. *Genomics* **76**, 110–116.
39. Aravin A.A., Sachidanandam R., Bourch'his D. *et al.* (20089). A piRNA pathway primed by individual transposons is linked to *de novo* DNA methylation in mice. *Mol Cell* **31**, 785–799.
40. Gunawardane L.S., Saito K., Nishida K.M. *et al.* (2007). A slicer-mediated mechanism for repeat-associated siRNA 5′ end formation in Drosophila. *Science* **315**, 1587–1590.
41. Cook H.A., Koppetsch B.S., Wu J. *et al.* (2004). The Drosophila SDE3 homolog armitage is required for oskar mRNA silencing and embryonic axis specification. *Cell* **116**, 817–829.
42. Qiao D., Zeeman A.M., Deng W. *et al.* (2002). Molecular characterization of hiwi, a human member of the piwi gene family whose overexpression is correlated to seminomas. *Oncogene* **21**, 3988–3999.

43. Taubert H., Greither T., Kaushal D. *et al.* (2007). Expression of the stem cell self-renewal gene Hiwi and risk of tumour-related death in patients with soft-tissue sarcoma. *Oncogene* **26**, 1098–1100.

44. Liu X., Sun Y., Guo J. *et al.* (2006). Expression of hiwi gene in human gastric cancer was associated with proliferation of cancer cells. *Int J Cancer* **118**, 1922–1929.

45. Grochola L.F., Greither T., Taubert H. *et al.* (2008). The stem cell-associated Hiwi gene in human adenocarcinoma of the pancreas: Expression and risk of tumour-related death. *Br J Cancer* **99**, 1083–1088.

Index

www.ingramcontent.com/pod-product-compliance
Lightning Source LLC
Chambersburg PA
CBHW050544190326
41458CB00007B/1911

9 781848 163669